Understanding Electronic Security Systems

By: M. Dean LaMont
Staff Consultant
Texas Instruments Learning Center

Managing Editor: Gerald Luecke
Mgr. Technical Products Development
Texas Instruments Learning Center

TEXAS INSTRUMENTS
INCORPORATED
P.O. BOX 225012, MS-54 • DALLAS, TEXAS 75265

This book was developed by:

The Staff of the Texas Instruments Learning Center
P.O. Box 225012, MS-54
Dallas, Texas 75265

For marketing and distribution inquire to:

James B. Allen
Marketing Manager
P.O. Box 225012, MS-54
Dallas, Texas 75265

Appreciation is expressed to Don Falk, Tim Shirey, Rose Mary and Dennis Banks, Steven Howard, Dave Woody, Waverley D. Cameron, Larry Tracy, and Miles (Bud) Sweeney for their valuable comments.

ISBN 0-89512-105-0
Library of Congress Catalog Number: 82-050800

A previous edition was published for Radio Shack in 1979 under the title *A SAFE HOUSE ELECTRONICALLY*

Table of Contents

Preface

Electronics systems are revolutionizing the world around us. Security systems are no exception.

Man has a basic need for security. He wants his family, his property – whether personal or business – and himself free of intrusion. This book provides information about how electronic devices and systems are being used to provide such security.

This book assumes no prior knowledge, except possibly a bit of basic electricity. Basic concepts and fundamental functions are stressed from the types of electronic detection to example applications.

The purpose of the book is to provide an understanding of the various types of detection systems that are available, how the systems work and the factors that must be considered when using electronic systems for security.

The example applications include using systems to provide personal security when occupants of the property are present and when they're not. Both home and business property examples are given. For the readers so inclined, more systems technical details including the need for computers, are presented in the last chapter.

Quizzes are provided at the end of each chapter to check how well one understands the chapter material.

Solid-state electronics, and integrated circuits in particular, have created high-performance, complex, low-cost, reliable, highly-portable products. This same sophisticated technology is now being used to satisfy man's need for security. This book is designed to provide an understanding of how this is being done.

Basic Security Needs

A COMMON SCENE

Picture a quiet April night. The palm trees sway softly. On the beach a slight wind blows the sand and stirs the grass growing by the road. The early spring sky is dark with clouds, and there is no moon.

At the far end of a gravel road, Sam Walker's beach house stands deserted *(Figure 1-1)*. Sam and his wife Laura have left their winter home once again and have returned North. A large van slowly approaches the beach road to Sam's house, turns in at the road and pulls around to the back of the house, being as quiet as possible. Two men, professional burglars, climb out. Having watched the Walker home for several days, they have observed that the neighbor down the beach drives up every few days to check on the house. They noted the drawn shades, the lack of daily mail and news. They also have watched the same light turn on and off every night at the same time. These are the signs of vacancy and these men stand ready to take off with a large share of Sam and Laura's belongings.

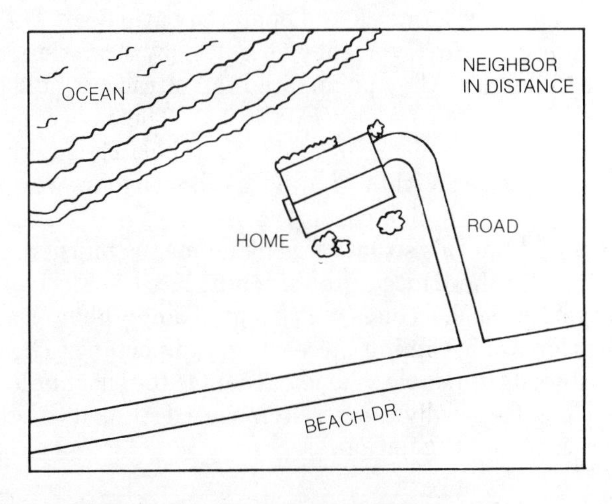

Figure 1-1. *Sam Walker's Beach House*

As professionals, they are aware of possible alarms. Their first act is to use a directional compass to locate a magnetic alarm switch attached on the back window closest to them. After covering the glass with masking tape, they open a fist sized hole in the glass and then place a small magnet against the switch holding it in the closed position. Next, they force open the window. They are ready to enter undetected and take all those things for which Sam and Laura have worked so hard for so many years.

Sam and Laura thought they had protected their home well. Sam had asked his neighbor to keep an eye on the house while he was gone for the summer. His neighbor agreed to visit every so often. He would see that the yard was picked up and everything was intact. In addition, Sam installed a lamp in his living area which turned on automatically at 7:30 p.m. and off at 10:30 p.m. Then last year, as an added precaution, Sam had a burglar alarm system installed. Have any of these actions prevented the robbery? No, they had not. Although Sam and Laura felt they had taken all possible steps to insure the safety of their home, it is obvious that they had not. Is there anything further they could have done? The average person would think not. However, let's back up again to the robbery. Only this time a different type of precaution has been added.

A FLASHBACK

After the burglars climb through the window, they walk into the living and den area, unlock and open the patio door. They begin to load up their van. However, they do not know that when they entered the living area they penetrated the ultrasonic field of a device which detects movement *(Figure 1-2)*. The device did not trigger an alarm but made a call to the neighbor. The neighbor, in turn, called the police. Within 15 minutes the burglars were arrested with a van half full of stolen property.

Sam and Laura's situation is by no means unusual. Every ten seconds a home in the United States is burglarized. Electronic detection systems such as ones which signal a neighbor or a district security station are becoming increasingly important to help satisfy the security needs of people who leave homes for short or long periods of time. Hopefully, such systems are designed to help reduce the crime rate across the nation.

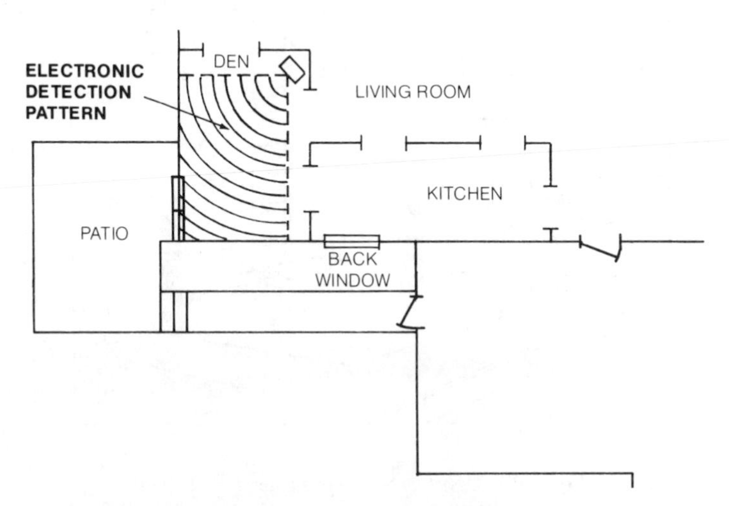

Figure 1-2. *Electronic Detection Field In Den Area*

ANOTHER COMMON SCENE

While it is important to feel that our homes are secure when we are away, isn't it even more important to feel safe while we are at home? Take the case of Susan Clark, an attractive, 24-year old, secretary for the production manager of a large corporation. It is 2:00 a.m. on a hot Sunday in July, and Susan is asleep in her suburban apartment *(Figure 1-3)*. She has taken several precautions to assure her safety. As she is aware of the possibility of rape, she does not drive alone late at night, nor does she attend any of the singles bars which abound in her area. She keeps her shades drawn and her doors locked. Although she is unaware of it, Susan has a problem. Bob Phillips, a young shop worker in Susan's corporation, has taken an interest in her. He has made several suggestive remarks to Susan and finally last week asked her out for a drink. Susan, not wishing to encourage this young married man, turned him down rather abruptly. In fact, she asked him to quit bothering her. Now on this hot summer night, Bob has decided that he will see her, and trying to bolster his courage, Bob has several drinks. When he arrives at Susan's door, he hesitates because he doesn't want to ring the bell; rather he will pick the cheap lock on her apartment door and force his way in. The story continues in much the same way as multitudes of like incidents that the newspapers carry each day.

Figure 1-3. *Suburban Apartment Building*

AGAIN, A FLASHBACK

But what if Susan had taken one added precaution? For less than $25 she could have purchased a small alarm box *(Figure 1-4)* to be placed on the inside doorknob of her apartment before retiring. Thus, at 2:30 a.m. when Bob touches the doorknob his body capacitance, connected to the doorknob, completes an electrical circuit sounding a loud buzzer and Bob flees. Susan is safe. What insurance! How many women living alone fear for their own safety each night? The cost of the electronic systems to provide security probably is a small price to pay for Susan's peace of mind. Is such a device on the market now, or is this story science fiction?

While it is true these stories are fiction, the electronic detection systems used as security devices are on the market now and at affordable prices. These up-to-date systems are designed and developed utilizing semiconductor integrated circuits, the latest state-of-the-art semiconductor components. The systems available range from simple closed-circuit hardwired systems to complex sytems that use ultrasonic, radio, microwave (radar) and infrared frequency signals to do their job.

The goal of this book is to help you understand the basic concepts of electronic detection systems and to show how they can be used to provide security for a person, his property and his country.

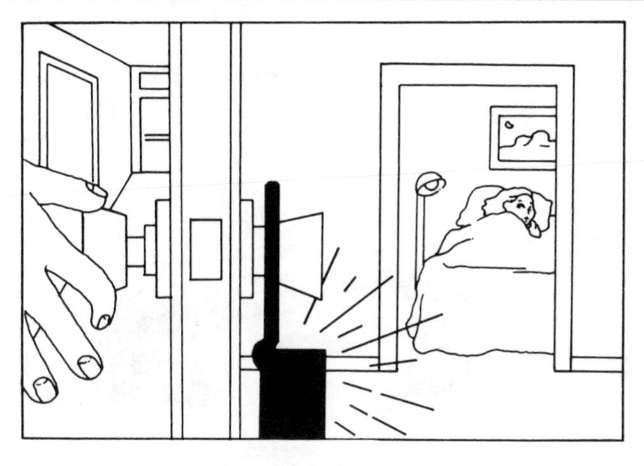

Figure 1-4. *Doorknob Capacitive Detection*

THE EVOLUTION OF SECURITY

Security is a basic need. Many philosophers and social scientists believe that is it impossible to live a full and rewarding life without a feeling of security. Early mankind undoubtedly had less security than we do today. A valid question to ask at this point is, "How did we get where we are today?" Are security needs really different than they were in the past?

A Historical Perspective

How did early man protect his belongings? Perhaps the biggest and the meanest took what they wanted? But no matter how big or mean, they still had their weaknesses; for instance, they had to sleep. And when they were asleep, even one who was small and meek could take a club and even the score *(Figure 1-5)*. Perhaps a better way to enjoy security was by being smart enough to outwit one's fellow man. By hiding most valuables and setting traps maybe one could be safe. A few crackly bushes placed outside one's sleeping quarters might be enough to alert one to a threat. But an alarm raised was of no value when the intrusion was made by a group of several against one. Several people together could always take one by force.

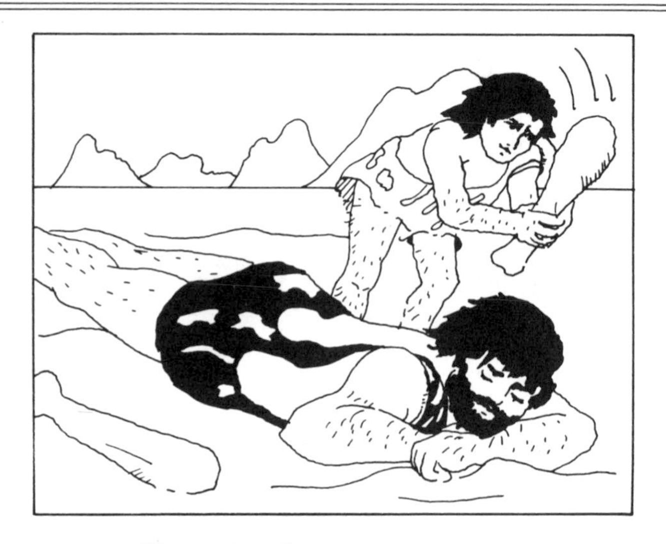

Figure 1-5. *The Small Evens The Score*

The Need for a Group

For many security reasons it became apparent that groups of individuals could more effectively handle the environment. A group could hunt larger game, and someone could always be left behind to protect the family and belongings. Perhaps the greatest material benefit derived from being in a group was that specialization of labor increased the amount of goods for all.

A Permanent Home

It is estimated that in early civilization the cultivation of previously wild plants encouraged the growth of permanent settlements. In addition, animals were domesticated as both a work and food source. The improvement in security was two fold. It was much easier to protect a permanent area and the food source was much more reliable. It was also no longer a necessity that all belongings be portable – heavy furniture and refined shelters became a possibility. However, as man grouped together *(Figure 1-6)*, coveting another man's belongings became a reality, but since these groups tended to be small and permanent, theft inside the group was minimal. What could one do with a stolen article anyway? Everything was handmade and easily identifiable.

Figure 1-6. *Picture Of Early Pilgrim Settlement*

The Door Lock

As civilization advanced, certain groups, tribes, nations, countries, and empires began to dominate others. Some dominated by finding particularly fertile locations. Other dominated by a common religion or political system which enabled them to unite their efforts in a common direction. Some were just more cruel, or more crafty, or just had more might. But as cultures grew, a new sort of threat became noticeable. Members of the same culture would steal the belongings of others, mostly from the rich or frugal. As a result, Egyptians as far back a 1000 B.C. found it necessary to use door locks. They used massive wooden bars in which wooden pins acted as bolts. The keys were about two feet (70 cm) long and weighed nearly four pounds (1.8 kg). The Greeks modified the lock with a different principle for the key and were able to reduce the key to about 12 inches (30.5 cm) *(Figure 1-7)*. Further modifications by the Romans reduced the size of the key to about six inches (15.2 cm).

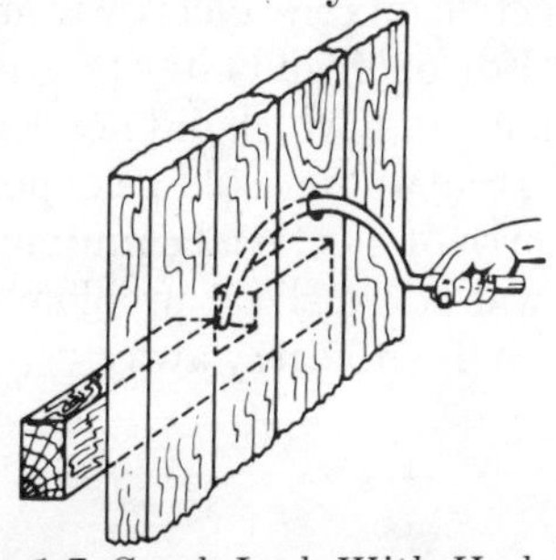

Figure 1-7. *Greek Lock With Hook Key*
(Source: Encyclopaedia Britannica, 14th Edition, 1958)

The Production Of Wealth

Around 1700 the Industrial Revolution began. Science paid off by providing new technologies. The Newcoman pump allowed water to be pumped out of coal mines. The coal was used to drive steam engines, which provided the power for industrial plants. Industrial plants drew people from widely scattered areas and with specialization of labor and new techniques material production soared. Cities grew as more and more people congregated in and around the industrial plants, despite the fact that work conditions were poor, hours were long and pay was meager. Even though the great gains in production were not passed on to workers in a fair share, the large bodies of people working in the plants were much better off than they had been or would be by working on the farm. With such a congregation of people, it was only human nature that people should want other people's property, especially the lowly worker who felt the rich industrialist was cheating him. The result was an increased crime rate and an increased feeling by man that he needed more individual security.

Thomas Edison's Electrical Gifts

After the American Civil War, Thomas Edison invented the electric light and the dynamo that began an electrical revolution, bringing comforts and luxuries to the home which were undreamed of in the past. Power from the dynamos ran the lights, motors and machinery for still greater industrialization *(Figure 1-8)*. One might say that Thomas Edison also gave us the beginning of our modern security problem and a boon to the criminal; for with the industrial revolution came the mass produced lower cost appliances now available to every worker. It became almost a certainty that every home in America had something which could be stolen and easily sold for quick cash: electric mixers, electric irons, phonographs (another invention of Edison's) and later televisions, radios, cameras and stereos. Now even a home which did not contain expensive silver or collectors' items of great value could be expected to yield several hundreds of dollars worth of electrical equipment. For the potential thief, however, there was still a slight problem – much of the equipment was rather bulky to carry away.

Figure 1-8. *Picture of Electrical Power Plant*

The Solid-State Revolution

With the advent of the transistor and later the introduction of the integrated circuit, electrical devices grew in complexity, as they shrank in size and weight. Literally thousands of transistors, diodes, capacitors, resistors and their required interconnections were placed on a single chip that can fit on a fingertip *(Figure 1-9)*. The integrated circuit provided outstanding performance in a very portable package. As a result, today's craze for good music almost dictates that a middle class home will contain several hundred dollars worth of stereo equipment alone – all packaged in a lightweight bundle for easy removal.

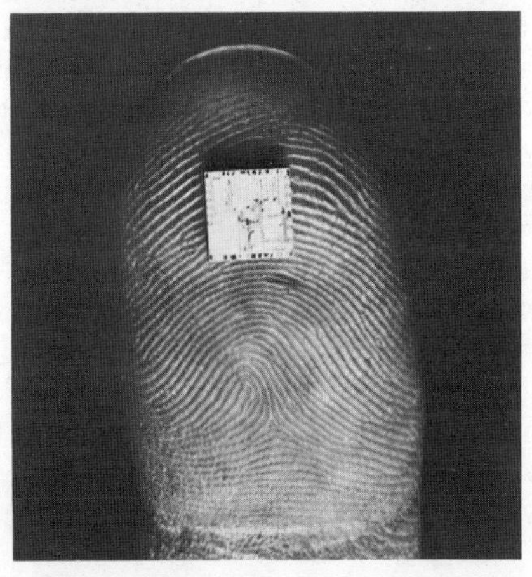

Figure 1-9. *Fingertip Electronics*

Add This In

Add to this situation various groups which are, or feel, discriminated against. People who feel that the system owes them something that they didn't get. People who feel they are free to take for themselves. Add to this the number of young people who do not make it through school. They are jobless or make only meager wages. They are engulfed in advertisements for this and that telling them how to be included with the "in" crowd. They can't afford what they think they need. They steal to be one of the crowd, for kicks, to feed a habit, or to live beyond their means.

Add in a highly mobile society where families have no roots and move every few years. Add in broken families where parents and children go their separate ways. Add to this the fact that many more people live in the city or its suburbs and the sum is a situation in which many persons feel alone and unknown. They feel they are just nameless faces in a crowd – and free to do as they please.

So society has come a full circle. Where man originally gave up his individuality to form groups for security reasons, he now is wanting to protect himself individually because he lives in close proximity to large numbers of people indifferent to his needs.

TODAY'S ANSWER

Get Away From It All

Get away from it all, that's one answer. Isolate yourself and your property from the crowd. Consequently, families have begun moving from the city to the suburbs and beyond to the country.

Because suburbia is where many people with above average incomes congregate, suburban areas which were relatively safe yesterday are, today, experiencing the greatest growth in crime. And the small towns and small cities associated with suburbia and the country are plagued with drastically increased crime.

Yet the increased crime rate doesn't concern other people. They feel their security needs are met. They have enough to eat, a roof over their heads, and enough material wealth to be comfortable. Their country is well protected with sophisticated missile systems so they have no worry. They have nothing of value that others would want and what they have is covered by insurance. Is there a flaw in this reasoning? As one would suspect, the answer is yes. One need only scan the newspaper and watch TV to verify the answer.

But I Have A Dog

For others the answer is a dog. A dog is certainly some help in alerting you to the presence of an intruder. The problem is, "old Fang" alerts his owner *(Figure 1-10)* to the presence of cats, squirrels, other dogs and possibly every passerby, so the effect wears off quickly. His barking begins to have the same effect as a snooze alarm on a clock.

Obviously dogs also have problems. Old Fang may have a weakness for tasty snacks such as meat containing a sedative, or age is just getting the best of him. At this point, the only alternative is to get a bona fide attack dog. When he arrives on the scene, the owner's popularity in the neighborhood declines as his dog food bill soars. And what happens when the neighbor's kid climbs over the fence to retrieve his tennis ball? He is faced with medical expenses or even a lawsuit.

Figure 1-10. *A Dog's Alert*

Oh Well, I Have A Gun

And still others favor a gun. When a gun is involved these questions naturally arise with regard to the owner:

How long has it been since the gun has been fired? Where is it exactly? Is it handy enough to get to in a matter of seconds? If awakened at night by a prowler *(Figure 1-11)*, who is most likely to be wide awake? A problem with bringing a gun into a confrontation is that it immediately escalates the encounter. If the intruder has a gun, he may now feel the need to use it. There are other possible problems involved with bringing a gun into the situation. Are you really prepared to use it? What if you shoot an unarmed burglar? There is a 50 percent chance he will be younger than 18. Will it traumatize your life?

Figure 1-11. *Awakened By A Burglar*

On the other hand, if the intruder is not a common burglar, but some sort of loony bent on violence, then a gun may be the best defense. That is for the owner to decide. The interest in this book is to provide other very viable alternatives

Oh Well, I've Got Good Insurance

For others it's insurance. If a large amount of property is lost, how much will the insurance company pay? Even if the original value of your property is reimbursed (which is unlikely), the owner will still be out hundreds or even thousands of dollars because of inflation.

Most insurance policies *(Figure 1-12)* require that proof of forced entry be shown. This means that the property must be locked when left alone, or a person may not have insurance at all. The loss due to malicious destruction of items such as photograph albums, keepsakes and momentos can be far more painful than the loss of a stereo or color television. Insurance is the last line of defense, or maybe the next-to-last line of defense – at least the loss (less $100 or whatever the deductible) is presently tax deductible.

Figure 1-12. *Insurance Policy*

I'll Just Take My Chances

And others will take their chances. Some people feel that the chances are just too low that their home will be bothered to justify taking action. After all, many of the neighboring homes may appear more desirable as a target than theirs. The spine chilling truth is that within a five year period one out of every five homes in America will be burglarized. Chance often determines who will be robbed. Most home robberies are committed by nonprofessional criminals, often juvenile, who are often on or in need of drugs. They set out to perform a particular type of crime, but the victim will often be determined by the whim of the moment. Ninety percent of today's burglars are under 35 years of age. They most commonly look for items that they can steal easily and sell quickly; stereos, cameras, jewelry, TV's and the like.

What Can Be Done To Increase Your Security

First of all, use some common sense *(Figure 1-13)*. Many of the criminals of today are not a hardworking sort. Any deterrent at all might be effective. Therefore, keep doors and windows locked (use the deadbolt locks the police department recommends), keep the area around the property well lighted, keep the neighbors informed so they know the status of the property and if someone is home or on vacation, keep the dog in different places, and be aware of any new people on the scene in the neighborhood. With these precautions it is very likely that the particular property will not be disturbed.

A) Keep Doors and Windows Locked (Use Dead Bolts)
B) Keep Areas Well Lighted
C) Keep Your Neighbors Aware of Status
D) Be Aware of New People on the Scene

Figure 1-13. *Common Sense Check-List*

Secondly, there are also some other very sensible and affordable alternatives now available. The same electronic technology which helped produce some of the escalating crime problems can also be used to combat it. The sophisticated technology which produced the portable wealth that's easy to steal may be just the means to protect it.

From the World War II era and into the space technology era significant scientific advances have been made to provide new means *(Figure 1-14)* for detecting the presence of unwanted persons. The advances in radio frequency transmission and reception are providing new wireless interconnection schemes and easier use of electronic detection systems.

1) Wireless Links
2) RADAR (Microwave)
3) SONAR
4) Ultrasonics
5) Doppler Effect
6) Seismographic and Geophysical Advances

Figure 1-14. *Technology Advances Contributing To Electronic Detection*

The advances in RADAR (Radio Detection and Ranging) and its associated technologies of transmitting energy at higher and higher frequencies has contributed to the development of microwave frequency electronic detection systems. These same radar technologies are providing new electronic detection systems that use sound waves in a similar way that they are used for SONAR (Sound Navigation and Ranging) systems. These systems may operate in the audible frequency range or in the above-audible (ultrasonic) frequency range. Advances that have been made in using the Doppler technique (discussed later) are making all of these systems practical for personal security.

Even the advances that have been made in seismographic and geophysical instrumentation for energy exploration and military security are resulting in electronic detection system components for security purposes.

WHAT'S NEXT

In the next chapter the various types of detection systems will be examined. Basic concepts are discussed, some application limitations identified, and advantages and disadvantages briefly summarized.

Quiz for Chapter 1

1. Early man's security was provided by:
   a. Being the biggest and the meanest.
   b. Being smart enough to outwit his fellow man.
   c. Gathering in groups.
   d. Isolating himself.
   e. a, b and d.
   f. a, b and c.
2. Industrialization and new technologies caused:
   a. People to migrate to the cities.
   b. A lowering of the crime rate.
   c. People to become independent.
   d. All of above.
   e. a and c above.
3. Crime statistics may be increasing due to:
   a. Discrimination.
   b. Crowded living conditions.
   c. Broken homes.
   d. Highly portable goods.
   e. Drugs.
   f. All of the above.
4. What % of today's burglars are under 35 years of age?
   a. 50%.
   b. 30%.
   c. 90%.
   d. 75%.
5. Most intruders select their target by:
   a. Careful planning.
   b. Chance.
   c. The yellow pages.
   d. A city map.
6. The most important common sense items for security include:
   a. Keeping doors and windows locked.
   b. Keeping areas well lighted.
   c. Keeping your neighbors aware of your status.
   d. Being aware of new people on the scene.
   e. All of the above.
7. Security needs today are different from those of the past because of:
   a. The growth of the city.
   b. The move to the suburbs.
   c. A highly mobile society.
   d. People indifferent to their neighbors.
   e. All of the above.
8. The common problems involved in using an attack dog for security are:
   a. All of the below.
   b. The food bill.
   c. Barking in non-security related situations.
   d. Danger to innocent victims.
   e. Possible lawsuits.
9. Common problems associated with using a gun for security are:
   a. A gun may escalate the encounter to a new level of danger.
   b. It may not be available when you need it.
   c. Children may accidentally discharge it.
   d. a, b and c above.
   e. It may be the wrong size.
   f. All of the above.

10. What % chance is there that an intruder is under 18?
- **a.** 50%.
- **b.** 65%.
- **c.** 20%.
- **d.** 5%.
- **e.** 80%.

11. Integrated circuits have grown in complexity such that one piece of semiconductor material contains:
- **a.** 1-10 components.
- **b.** Only the wires and connectors.
- **c.** Thousands of transistors, capacitors, diodes and resistors.
- **d.** Less than one hundred devices.
- **e.** None of the above.

12. The Solid-State Revolution:
- **a.** Was like the Industrial Revolution.
- **b.** Provides high-performance, highly portable, desirable items in a small package.
- **c.** Doesn't have anything to do with electronics.
- **d.** Was over too soon.

13. Using common sense to help provide security is successful because:
- **a.** Everyone thinks alike.
- **b.** A penny saved is a penny earned.
- **c.** Many intruders of today are not a hardworking sort.
- **d.** A gun is all you need.
- **e.** No one will bother a locked house.

14. Insurance is only partial protection because:
- **a.** It is difficult to set the rates.
- **b.** Photo albums, keepsakes and mementoes cannot be replaced.
- **c.** The company may not remember your policy.
- **d.** The insurance companies are too particular.

15. Radar techniques have contributed to the development of:
- **a.** Hardwired perimeter systems.
- **b.** Wireless perimeter systems.
- **c.** Microwave frequency electronic detection systems.
- **d.** Infrared systems.

16. SONAR which works on the same principle as RADAR, means:
- **a.** Source of noise area.
- **b.** Sounding around.
- **c.** Source of navigation route.
- **d.** Sound navigation and ranging.

17. The same electronics technology that has produced some of the crime problems:
- **a.** Can also be used to combat crime.
- **b.** Provides products that are very high cost for the average consumer.
- **c.** Has had no impact whatsoever.
- **d.** Is going out of style.

18. Advances in radio frequency (rf) transmission and reception has contributed to:
- **a.** Wireless detection systems.
- **b.** Hardwired systems.
- **c.** New magnetic switches.
- **d.** All of the above.

(Answers in back of the book)

The Types of Electronic Detection

ABOUT THIS CHAPTER

Since security is such a basic human need, how can electronic systems help to provide security? What means are available and what are the basic concepts used? This chapter provides such an overview and introduces such terms as hardwired, wireless, area and space detection, infrared, microwave, etc.

GENERAL DETECTION SYSTEMS

Security systems in general can be shown to provide the three basic functions of detect, communicate, and act, as shown in *Figure 2-1*. As an example, a dog detects that a person is an intruder by sight, smell, hearing and touch. His senses communicate this fact to his brain and he acts by barking and possibly biting. Electronic watchdogs also detect, communicate, and act. Usually the action at least sounds an alarm, but it can be more extensive, such as providing signals to remote locations.

Figure 2-1. *Basic Detection System*

Recall in the early days that warnings were provided to people through gunshots, church bells, factory whistles, boat whistles, beacon lights and sirens. Early settlers raised their guns and fired shots into the air to signal they were in trouble. Large ropes were pulled to make the bells toll to sound an alarm. Steam pressure from boilers provided the power source for the factory and boat whistle. An electric circuit, as shown in *Figure 2-2*, where an electric motor received power when a switch was thrown, was used for the siren alarm. Today flashing lights and sirens and radio and TV transmissions are the main avenues of warning or sounding an alarm.

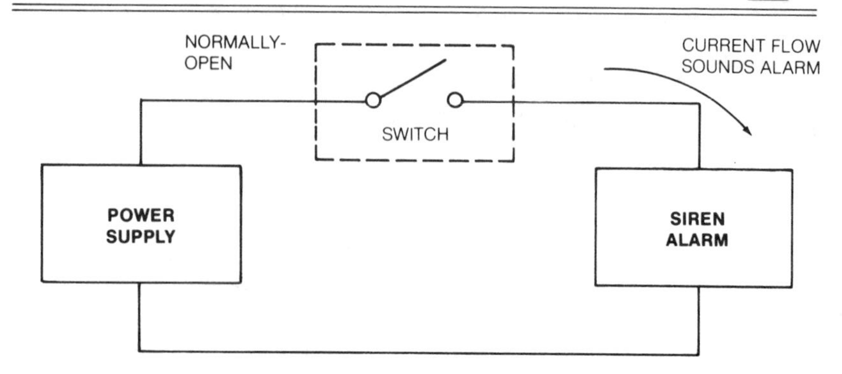

Figure 2-2. *Electric Circuit For Siren Alarm*

BASIC DETECTION SYSTEMS

Most of the simpler electronic detection systems are based on an electric circuit similar to that of *Figure 2-2.* They basically consist of a power source, a wire connection to complete the electronic circuit, a switch, and some form of alarm. The power source may be an alternating current source derived from the one supplying the 110-220 volts power service to a home or business; or it may be a direct current power supply which rectifies an alternating current; or it may be a self-contained direct current source such as a battery. In some cases it may be a combination, such as a direct current power supply and a battery back-up in case the public utility power source fails.

Normally-Open Switch Systems

The type of switch used in most basic systems is the normally-open type shown in *Figure 2-2.* The alarm circuit conducts current, dissipates power, and sounds the alarm only when the switch is closed. Simple alarm systems such as this are available for many applications – lighting lights, ringing bells, sounding alarms when doors or windows are being opened (or sometimes closed). A very common application is in automobiles to indicate that oil pressure is low or high-beam lights are on. The types of switches available include push button switches, trap switches, plunger type switches (similar to the ones which turn on the interior lights of your car when a door is opened), toggle switches (similar to your home light switches), magnetic switches and many others. The different types of switches will be described as the discussion continues since they are used in both simple and perimeter alarms.

One prime example of the use of a toggle switch, or push button switch, or knife-contact switch in a normally-open circuit such as *Figure 2-2* is the fire alarm contained in buildings and on street corners. To sound the alarm (it used to be that you had to break the glass first) a button must be pushed, or a toggle switch thrown or a knife contact lever switch pulled. These types of switches are shown in *Figure 2-3*. The normally-open switch is closed and the siren sounds the alarm. It might be noted that after all of these switches are closed they remain closed. Some push-button and toggle switches are momentary and return to the open position if released.

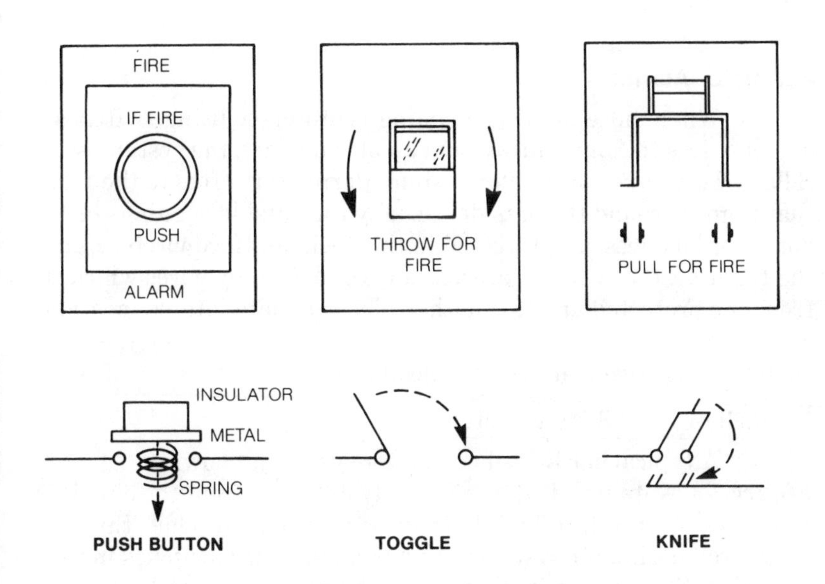

Figure 2-3. *Types Of Normally-Open Switches*

The basic simple alarm circuit can be extended to protect several openings (doors, windows, etc.) by adding more normally-open switches across each other (in parallel) as shown in *Figure 2-4*. If any one of the normally-open switches is closed, current flows and the alarm sounds.

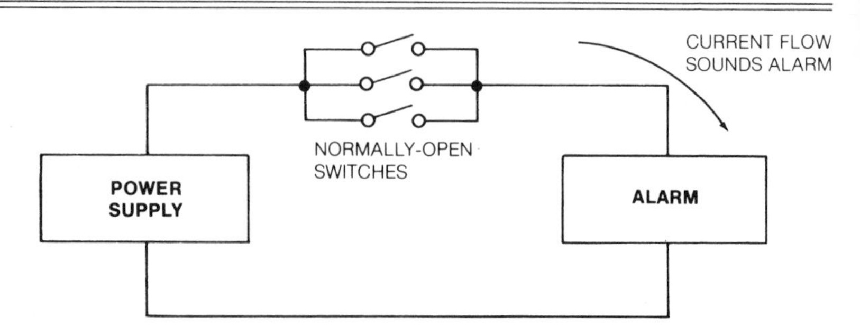

Figure 2-4. *A Simple Alarm Which Protects 3 Openings*

Perimeter Alarms

When more than one opening is protected using switches to detect, wires to communicate and an alarm to act, the system is called a *hardwired perimeter* system. Perimeter refers to the fact that it goes around the outside. Usually the outside openings of homes or business are protected in this fashion. Hardwired refers to the fact that the switches are actually connected by wires which run from one protected area to another. The complete interconnection circuit path for current flow is provided by wires between the switches, the power source and the alarm.

Normally-Closed Switch System

A typical hardwired perimeter system is shown in *Figure 2-5.* Note that it is significantly different from the circuit in *Figure 2-2* or in *2-4.* It has all of the switches closed. Thus, this is a hardwired perimeter system with normally-closed switches in the current path. The circuit symbols for the normally-closed version of the same switches shown in *Figure 2-3* are shown in *Figure 2-6.* Such switches installed in the system of *Figure 2-5* will be normally closed but open when disturbed. Therefore, in the circuit connection of *Figure 2-5,* they are providing a path for current flow around the alarm – they are in parallel with the alarm. As long as each switch is closed all the current in the circuit flows through the path with the normally-closed switches rather than through the alarm. Once any one of these switches is disturbed, it opens, breaks the parallel current path and forces the current to flow in the path through the alarm, sounding the alarm. A resistor is used in the circuit to keep from short-circuiting the battery when the switches are closed.

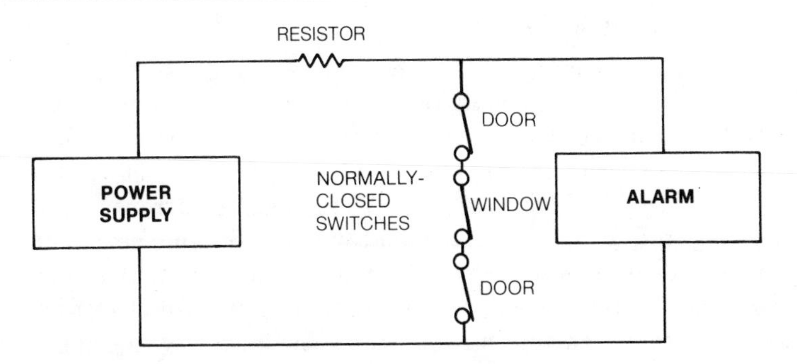

Figure 2-5. *An Alarm Circuit Using Normally-Closed Switches*

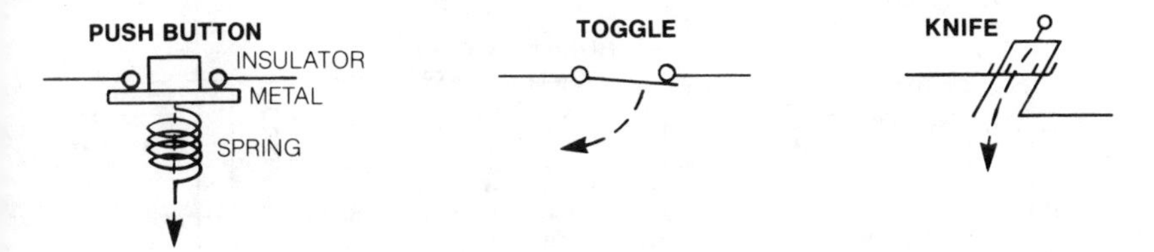

Figure 2-6. *Normally-Closed Switches*

If any of the wires interconnecting the switches are cut or disconnected the alarm sounds. This gives a two-fold advantage: If a wire is pulled loose accidentally during cleaning the alarm will sound as soon as the system is activated, and the problem can be rectified. Also, if a would be intruder cuts the wire the alarm sounds. (In a simple open-circuit system a damaged or cut wire would not sound the alarm.)

There are some disadvantages with a simple system of this type. First, current is continually drawn from the battery. With only a resistor for adjustment, this current can be excessive, thus wasting a great deal of power. Second, the alarm turns-off as soon as the switch is closed again. An intruder could break in, turn-on the alarm by opening a door, shut the door, turn off the alarm and proceed.

Combination Switch System

However, sometimes it is best not to build a system with just normally-open or just normally-closed switches. A combination is required. For this reason perimeter systems have more complex circuits so that both normally-open and normally-closed switches can be used. Solid-state devices are used to reduce the current drain on the battery and to hold the alarm on for a specific time period once the line has been broken (for the normally-closed branch) or once it has been closed (for normally-open branch). A general block diagram of such a system is shown in *Figure 2-7.* The normally-closed branch might be used around windows and doors and the normally-open branch for pressure-sensitive or panic buttons.

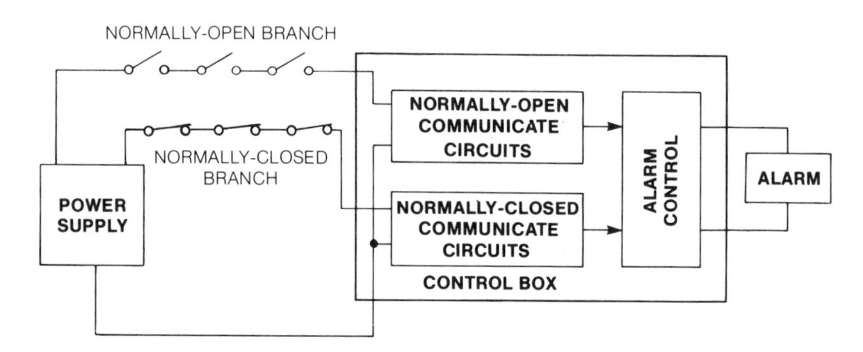

Figure 2-7. *Complex Perimeter System*

Protective Switches

Many types of protective switches are available for hardwired systems. For doors and windows, magnetic and plunger type switches are available (normally-closed most commonly used). Walkways may be protected by pressure-sensitive switches (usually normally-open) placed under a carpet or in a mat. Heat-sensitive switches (also usually normally-open) may be placed in furnace or fireplace areas. Mercury switches which detect a change in position may be used on garage and cellar doors. Switches which detect vibration may be placed in critical locations to detect entry or movement.

Doors and windows need not have switches attached to them as shown in *Figure 2-8*. They can be an integral part of the connecting wire system by having a thin conducting strip of metal attached to the glass which completes the conducting loop. This current path is the same as the normally-closed switch path of *Figure 2-5*. If the glass is broken, breaking the metal strip, the alarm sounds.

Figure 2-8. *Store-Front Detection System*

Water sensitive switches may be used to detect a flooding bathroom or basement. Pushbutton switches (panic buttons) may be placed in key locations through the home so that the alarm can be sounded by merely pushing a button. Thus, a hardwired alarm system can become a home control center to sound the alert in case of burglary, fire, flood or panic.

Wireless Perimeter Systems

The hardwired perimeter system just discussed has one major disadvantage – it is often difficult to install the interconnecting wires. This is especially true if the system is installed into existing buildings. It is not quite so difficult when the building is under construction. The aim of the wireless system is to overcome this difficulty because it is much easier to install, especially in existing buildings. As shown in *Figure 2-9*, wireless refers to the fact that the *communicate* portion of this system is done by radio waves rather than by wires as in the hardwired systems. At the time that the switches are installed, a small transmitter (about the size of a cigarette package) is connected to each switch. As shown in *Figure 2-10*, when the switch is triggered, the transmitter functions similarly to those used to open garage doors automatically. It generates a radio frequency signal which travels to a receiver to generate the alarm. The transmitted signal may be in the frequency range from 88 MHz to 108 MHz.

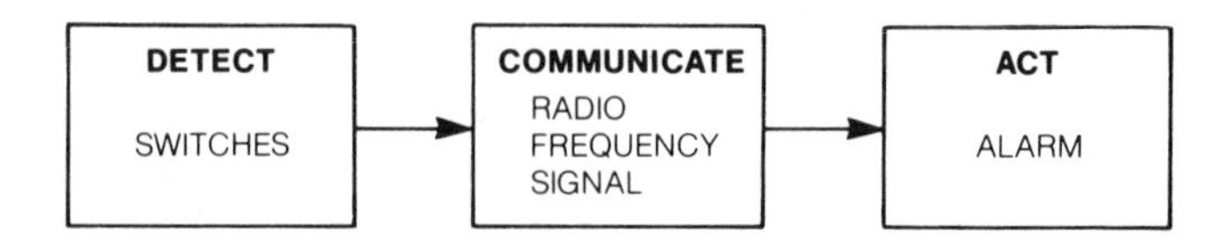

Figure 2-9. *Functions Of A Wireless Perimeter System*

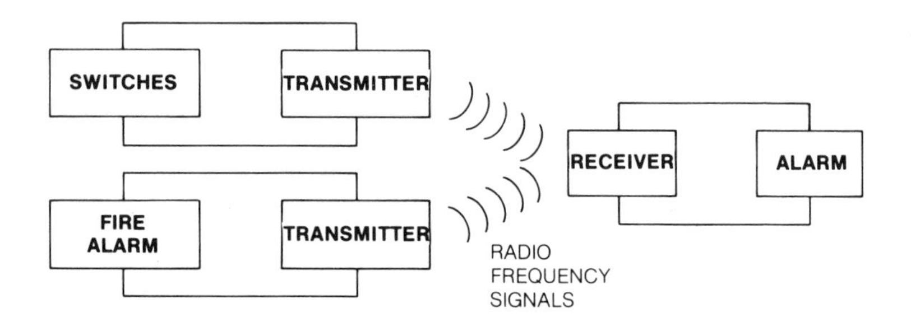

Figure 2-10. *A Wireless System Diagram*

Combination systems are very possible. For example, hardwired and wireless can be mixed by connecting more than one switch with each transmitter. This allows the windows in one room or in one area to be hardwired together. The signal from the transmitters is usually coded in one of two ways as shown in *Figure 2-11*. The first *(Figure 2-11a)*, employed by single-code radio controls, involves the modulation of the rf carrier signal by a very high frequency tone above the audio range. The second coding scheme *(Figure 2-11b)* uses a digital technique where the high-frequency carrier is switched on and off in a distinctive pattern of pulses. Both coding schemes work fine in the absence of interference. However, digital coding offers far greater assurance that the alarm won't be triggered in response to an alien signal. In addition, as more complex computerized systems evolve, coding the various transmitters differently will help to identify the signal area.

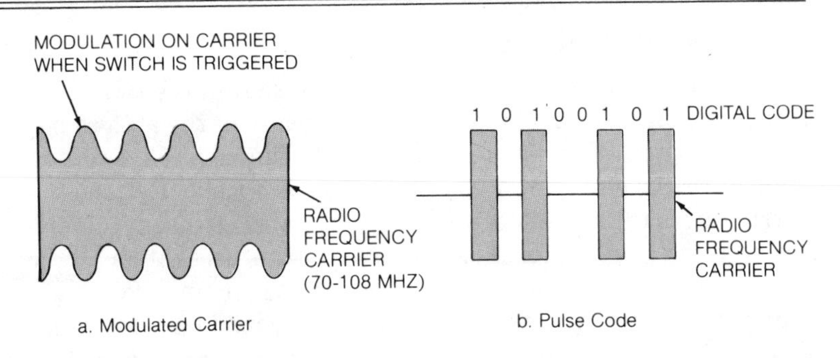

Figure 2-11. *Wireless Signal Coding*

Wireless systems offer several advantages in addition to easy installation. Portable transmitters may be carried or placed in strategic locations around the protected area. They may be particularly advantageous in an emergency. The alarm is sounded when a switch is pressed that is connected to one of the portable transmitters. This feature may be extremely valuable to those suffering from ill health. If a heart attack were to occur, paralysis or some other form of seizure, this would be a convenient way to sound the alarm.

Other benefits may occur. For example, the signal will often carry over 100 feet (30.5M), making protection of sheds and outer buildings convenient. Smoke detectors which transmit the same coded signal are often available with wireless units so the system can also provide fire protection. In many cases the individual smoke detectors in common use have a weak alarm volume. No matter how large the protected area, a wireless smoke detector that sounds the main system central alarm is certainly going to be heard.

Obviously, since transmitters must be added to the switches (or some combination of switches) and since receivers must be provided, there will be additional expense when the wireless system is used over the hardwired system.

A Variation Of The Wireless System

Wireless systems can be designed so that the radio frequency signal uses the power lines already in the area to be protected as the distribution network. As shown in *Figure 2-12*, each transmitter is plugged into an existing electrical outlet and when it is triggered, the radio frequency signal travels to the receiver through the power lines. Since the rf signal is present on all the power wires, remote alarm units may be used in any electrical outlet. Anyone causing the alarm to be triggered would have a tough time discovering and destroying the receiving unit when sound surrounds him.

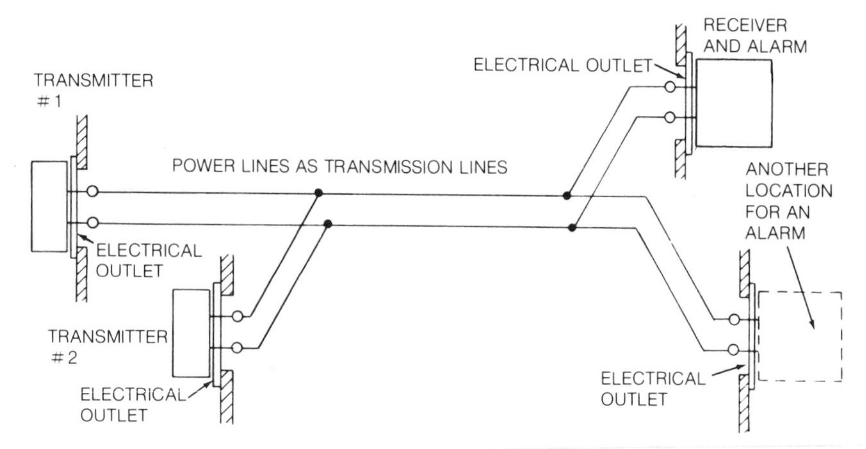

Figure 2-12. *Power Line Distribution*

Installation of a wireless unit which uses power lines for signal transmission can be a problem. Here are some disadvantages besides the additional cost:

1) It is difficult to predict the distance that the signal will travel on the line.
2) The number and kind of devices plugged into the power line vary the performance.
3) The signal will not travel between outlets which are supplied power from a separate utility company transformer.
4) There could be an rf interference problem with other appliances in the house.

These problems apply more to industrial installations than residential.

DETECTION BY LISTENING FOR SOUNDS

A Typical System

Basically the systems described thus far are very similar but the interconnecting loop has been the difference. Now, however, the basic media used for electronic detection changes. The first new media uses sound waves, in the specific case noises not normally in the area. A typical system is shown in *Figure 2-13.*

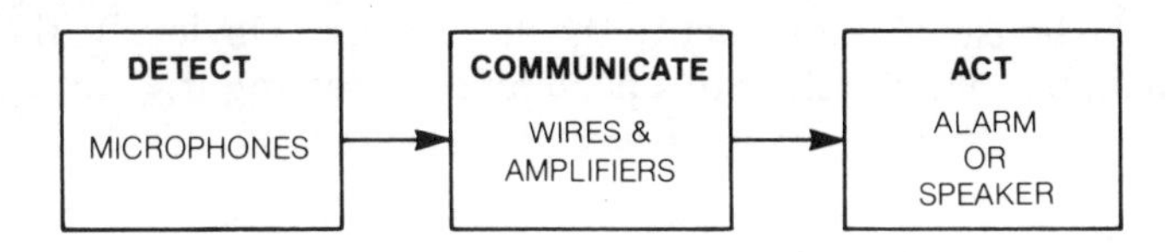

Figure 2-13. *Functions Of An Audio System*

The central unit for detection is the microphone, the central unit for communications is an audio amplifier and interconnecting wires, and the central unit for the action is a speaker or alarm.

Picking locks makes noise, forced entry makes noise and rummaging through drawers for jewels, money, and other valuables makes noise. A noise detection system is essentially a listening post that uses a microphone to put an intruder "on the air" where his every move can be heard.

Home Intercoms

One very simple way of accomplishing such a system of detection is to use the intercom system that has been installed in a home. Placing the intercom system on *listen* for each room, the speakers become microphones, the intercom amplifier the communication unit, and the one room speaker that is left on *intercom* becomes the output speaker. In this manner one location can monitor the sounds in all the other rooms and the total system becomes a sound detection system.

Independent System

As shown in *Figure 2-14*, an independent system is created by placing multiple microphones in the area to be protected, interconnecting them to a central amplifier and connecting a speaker to the amplifier's output. Disturbances at any microphone provide input signals to the central amplifier which are "heard" through the speaker. The speaker can be placed in a critical location so that a person can react to the sound in an orderly fashion. For example, a good place might be next to the alarm clock so that the head of the household can react to the disturbance. Obviously, the noise could be very startling to a person not expecting it.

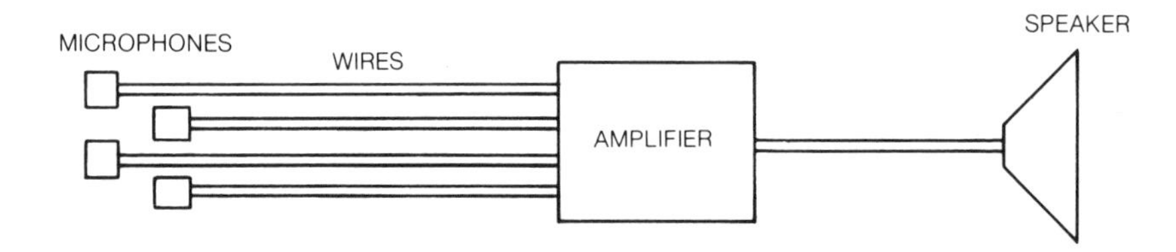

Figure 2-14. *Diagram Of Sound Detection By Listening*

Such a system can be distributed quite easily through some other systems that already exist. There might be an intercom system between buildings in a business location. The intercom microphones could be "opened" to receive the disturbance signal from a sound detection system installed in each building. In this way one security guard can monitor all the buildings. If a person is not available to listen at all times, there are electronic systems available that will do the monitoring and sound alarms when a preset level of sound is heard.

Differential Amplifier Systems

Sound detection systems have the advantage of being relatively inexpensive and fairly easy to install. Obviously, one of the big disadvantages is false triggering or detecting and amplifying sounds that are not security related. In order to solve this problem, differential amplifiers are used as the communications amplifier. The differential amplifier "subtracts" signal inputs from two microphones before amplifying. Suppose, for example, that two microphones are placed in separate rooms A and B as shown in *Figure 2-15*, and are connected to a differential amplifier and an alarm. Now imagine that a storm is in progress and there has just been a loud clap of thunder. The sound enters both microphones A and B *(Figure 2-15a)* so the electrical signals from the two microphones are both at a high level. Subtracting level A from level B gives approximately zero signal to be amplified, so the alarm is not sounded. A common signal has been received on each input, therefore, the differences between the signals is zero and the output of the differential amplifier is zero. This is called the common-mode rejection capability of the differential amplifier.

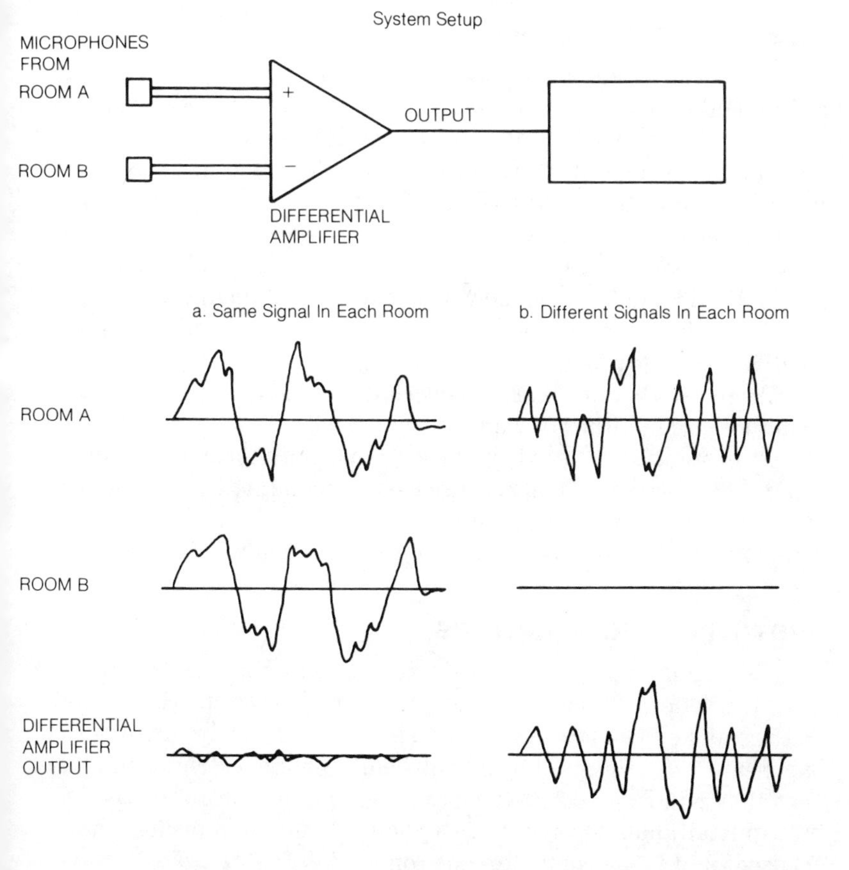

Figure 2-15. *Electronic Sound Detection With Differential Amplifiers*

On the other hand, suppose that someone breaks into the window of room B. The noise will generate a much stronger signal in microphone B than in A *(Figure 2-15b)*. The difference will be considerable and, when amplified, will sound the alarm. In addition, these systems also often use time sampling circuitry. This means that the sound is examined over a period of time before the alarm is sounded. It means that the sound levels in room A and B would have to be different for several seconds before the alarm would sound giving further protection against false alarms. Differential amplifiers need two inputs. Adding microphones by two's will require a differential amplifier for each pair.

Further System Refinements

Because sound is a good detection means, several significant refinements are available in sound detection systems. Since much of the noise that's heard is not really relevant to security, steps have been taken to make such systems susceptible only to the sounds that identify danger to security. Shattering glass, cracking wood, the human voice, they all have distinct acoustic spectrums – that means they have clearly identified sound frequencies that make up the total sound. For example, the male voice generates pronounced peaks at the 200 to 500 Hz (cycles per second) range. In audio units of this sort electrical signals produced by the sound waves in the microphones are filtered so that only certain selected frequencies can enter the amplifier circuit, other unwanted frequencies are rejected. When the electrical signals in the selected ranges are high the alarm sounds. In other words, these units discriminate between intrusion noises and noises which are of no interest with regard to security purposes. False alarms due to unimportant sounds are held to a minimum with these systems.

PHOTOELECTRIC DETECTORS

A common experience as one enters a shop is as follows: you open the door and you hear a buzzer, or a bell, or a tone that signals to the shopkeeper the entry of a customer. Most likely the tone that was heard was triggered by a photocell. A photoelectric system like that shown in *Figure 2-16* consists of a focused light source as the transmitter and a receiver with a photocell mounted inside. The focused light beam completes the connection for the system.

Whenever someone walks through the beam or the beam is interrupted in some manner, the light reaching the photocell is reduced causing a current fluctuation which is detected and amplified and triggers a circuit that sounds the buzzer.

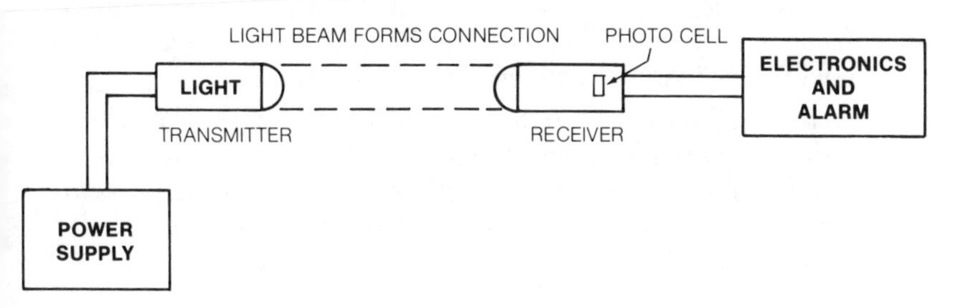

Figure 2-16. *A Photoelectric Detector*

Visible Light System

The photoelectric cell is sensitive to light rays (photons of light) and it responds by producing a small electrical change which is detected and used to trigger the alarm. This small electrical charge is a voltage if a silicon photocell is used, or it is a change in resistance in the detection circuit if it is a cadmium sulfide cell. These are called visible light detectors.

Visible light is satisfactory to detect daylight entry into a store, but it is not well suited for an alarm against a nightime intruder. All the intruder would have to do to avoid detection is shine a flashlight into the receiver as he passed through the beam. A solution to the problem is to use infrared light which cannot be seen by the human eye, because it is a different frequency than visible light.

Infrared Systems

One way to generate an infrared beam is to place the proper optical material in front of the standard incandescent light source so that the visible portion of the light is blocked leaving only the infrared portion to pass on through. Such optical lens arrangements can be used to focus the beam to cover a small area. One other very common source of infrared is a gallium arsenide diode. This is a special kind of semiconductor material that can be used to make devices that operate at infrared frequencies.

"Hot bodies" such as lighted cigarettes, matches, glowing embers, are sources of large amounts of infrared radiation. In fact, the infrared radiation from all bodies is proportional to the temperature of the body. This can be used as a means of electronic detection and will be discussed later. However, the "hot body" radiation can also be a disadvantage because a cigarette or lighted match could be used by an intruder to simulate an infrared beam system while the intruder slipped through it.

To overcome this, the infrared beam sometimes is pulsed electronically to further reduce the possibility of an intruder fooling the system. Not only can the beam not be seen, the coding prevents it from being simulated using other infrared sources – an intruder would find it very difficult to produce a properly pulsed infrared signal.

Perimeter Systems

Infrared energy is a form of light and can be reflected by mirrors. If the energy beam is strong enough it can be reflected several times to form a sort of invisible perimeter system *(Figure 2-17a)*. A significant advantage for infrared systems of this sort is that they are very stable and not prone to false alarms. Such a system is very desirable when an open path can be located for they are reliable, long-lasting systems that can be maintenance free. There may be a problem in some industrial situations where dust and dirt may block the path of the beam. As with any light system, false alarms might occur. Obviously, if any of the reflecting mirrors go out of adjustment the alarm will sound. A transmitter and receiver are shown in *Figure 2-17b*.

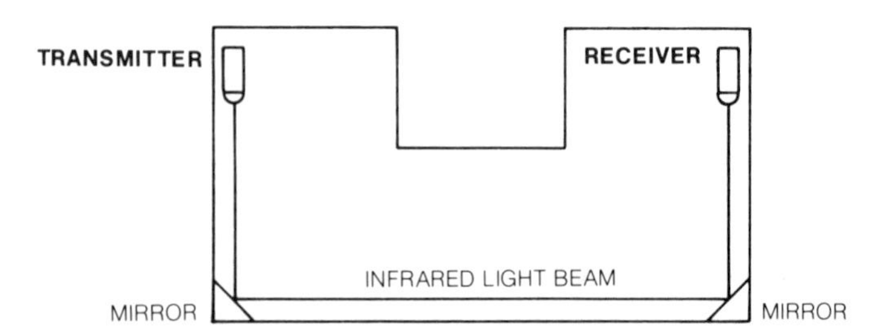

Figure 2-17a. *Infrared Perimeter System*

Figure 2-17b. Infrared Transmitter and Receiver
(Courtesy Radio Shack)

SPACE PROTECTION

The discussion now shifts from systems that use wires and switches and beams of light that are forming paths to be opened or closed to systems that detect radiation patterns that cover complete rooms. These are called space systems because they form a detection system for a certain amount of space. For example, look at *Figure 2-18.* A transmitting source is radiating a pattern of energy. Any object within that pattern is a possible target for detection. Any object outside the pattern will not be detected. Therefore, the space within the detection envelope is that which can be "protected" by the space detection system.

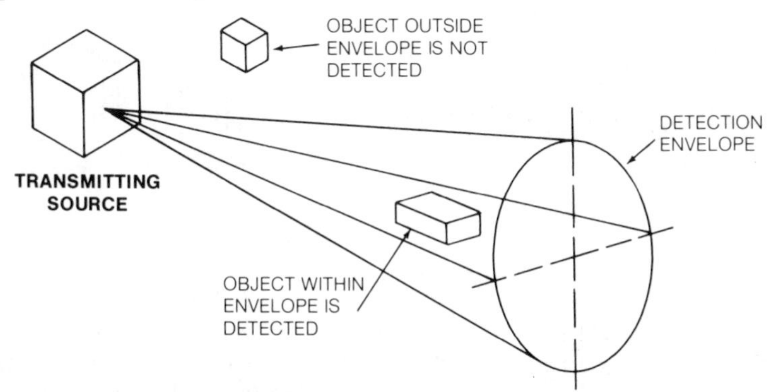

Figure 2-18. *Space Detection Envelope*

RADAR Systems

These systems are derived from a principle used in radio detection and ranging (RADAR) systems or from SONAR (Sound Navigation And Ranging) systems. Let's look at RADAR principles first, since such systems were developed before SONAR.

As shown in *Figure 2-19a*, a transmitter is sending out a narrow beam of very high-frequency energy (gigahertz – billions of cycles per second). It is not a continuous signal but comes in very short bursts or pulses spaced apart in time as shown in *Figure 2-19b*. Once the high energy pulse is transmitted, the system waits to detect any energy that might be reflected from a target.

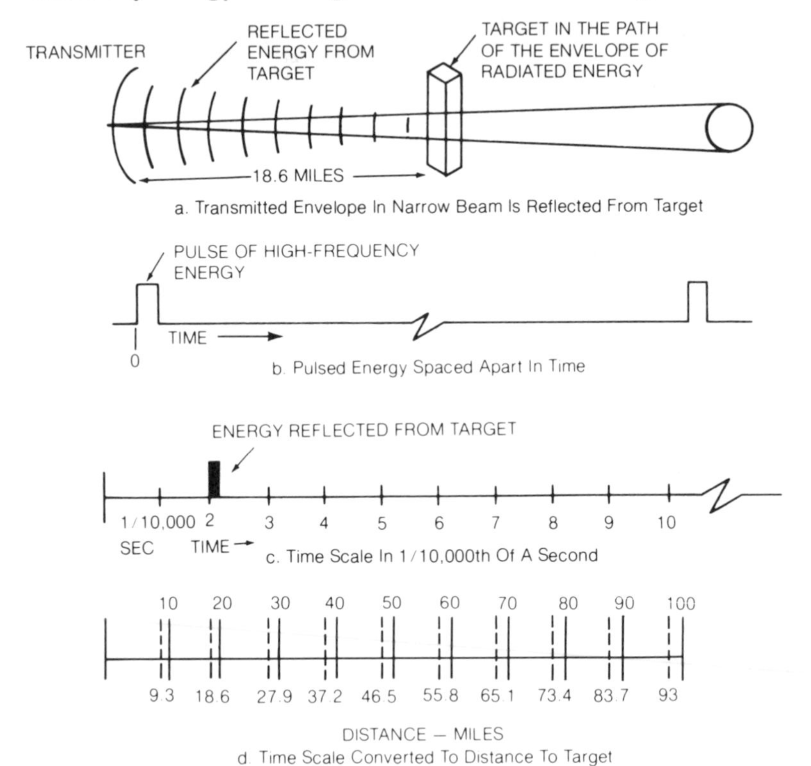

Figure 2-19. *Radar Principles*

A target is shown in *Figure 2-19a* 18.6 miles away from the transmitter. Electromagnetic energy such as contained in the pulse of energy from the transmitter travels 186,000 miles per second. It takes 1/10,000th of a second to travel to the target and 1/10,000th of a second to travel back to the location of the transmitter. Therefore, if a time clock is started at the time that the transmitter pulsed, a reflected pulse of energy should be detected by a receiver back at the transmitter at 2/10,000th of a second. This reflected pulse of energy is shown on the time scale of *Figure 2-19c*.

Now, the time scale of *Figure 2-19c* can be converted to a distance scale of *Figure 2-19d* that shows the distance to any target that might be received as a result of a pulse that was sent from the transmitter. The target of *Figure 2-19c* appears at 18.6 miles on the distance scale of *Figure 2-19d*. A more typical division of the distance in miles is shown on the 10, 20, 30 etc. scale superimposed on *Figure 2-19d*. Thus, any target on this scale could be detected in position up to 100 miles away. Local TV weather forecasters normally display radar scales from 50 to 400 miles and many surveillance radars such as shown in *Figure 2-20* may have ranges up to 1,000 miles.

Figure 2-20. *Picture Of Radar Surveillance System*

The time scale is normally displayed on a cathode ray tube screen and the time scale is the illuminated beam which is synchronized to the position of the antenna. As the antenna rotates in a 360° scan the time-scale beam travels with it. Any targets received as a result of a transmitted pulse appear as bright spots on the calibrated illuminated beam. Thus, the whole pattern of bright targets appear on the screen in their relative position and distance from the transmitter.

SONAR Systems

The same principles apply to SONAR systems as for RADAR. The only difference is that the high-energy pulses are now at the frequency of sound and reflected waves can be heard and detected as sound signals. SONAR systems were originally developed to operate underwater because the RADAR signals would not transmit through water – they were all absorbed. Therefore, based upon the principle of reflected energy from pulsed transmission two systems have emerged – one a very high frequency system and one a low frequency system.

These same principles are now being applied to space detection systems for protection against intrusion. Further sophistication has occurred to make the systems particularly suited for this application.

Doppler Systems

Units are available which transmit a wave pattern of a particular frequency into the space needing protection *(Figure 2-21)*. When the waves strike an object in the room or a wall, some of the energy is reflected back to the transmitting source. The waves which are reflected back have the same frequency as the generated waves (the distance between the waves stays the same). But, if an object is moving toward the transmitter and reflects energy back, the frequency of the reflected wave is slightly higher than the transmitted wave (the distance between the waves is reduced), and if the object is moving away from the transmitter, the reflected waves are slightly lower in frequency (the distance between the waves is increased).

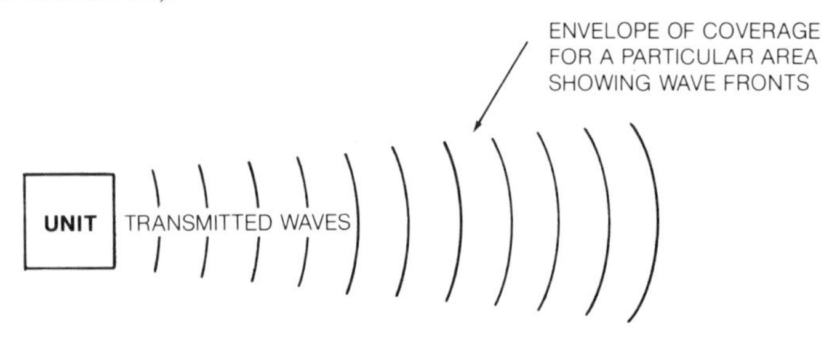

Figure 2-21. *Space Detection Unit*

This effect is called the Doppler shift and is illustrated for a person moving toward the unit in *Figure 2-22.*

When the transmitted wave 2 reflects from the person, he is located in the position shown, but when the transmitted wave 1 reaches the person he has moved toward the unit, so reflected waves 2 and 1 are closer together than when they were transmitted. Complex electronics in the unit detect this frequency shift and sound the alarm.

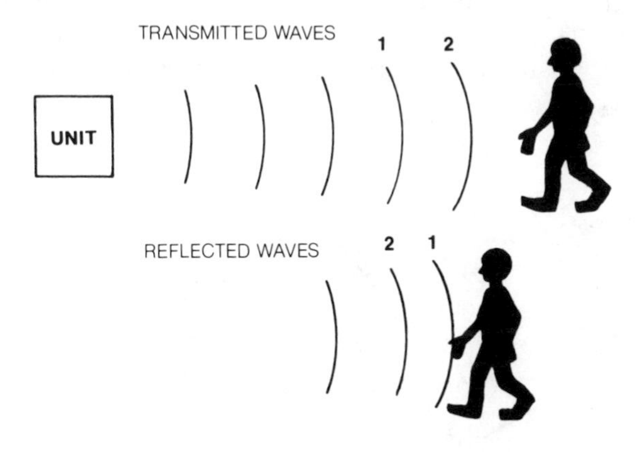

Figure 2-22. *Doppler Shift*

The Doppler effect is further illustrated by a speeding object coming toward a person with its horn or whistle blowing. A train is a prime example. As the speeding object approaches, the frequency of the sound from the horn increases and after it passes the frequency of the sound decreases.

Thus, space detection systems such as this, even though their energy envelope hits an object and is reflected and detected, it is only the moving targets that cause the alarm to sound.

Space detection systems operate at a number of different frequencies. The first of these is at a frequency just above sound, therefore, it is called an ultrasonic system.

Ultrasonic Space Detectors

Many ultrasonic systems resemble a stereo speaker in appearance (*See Figure 3-3*). As a stereo speaker sends out sound waves which are reflected about the room, an ultrasonic detector sends out sound waves which are above the frequency of human hearing (*Figure 2-23*). This energy travels out into the area in an elliptical pattern. The pattern is automatically adjusted on some units while others allow one to select a range from about 12 to 35 feet. The frequency of ultrasonic energy transmitted for detection is 20 to 50 kilohertz, therefore, ultrasonic waves do not penetrate through windows and walls so the field of protection is normally contained in one room.

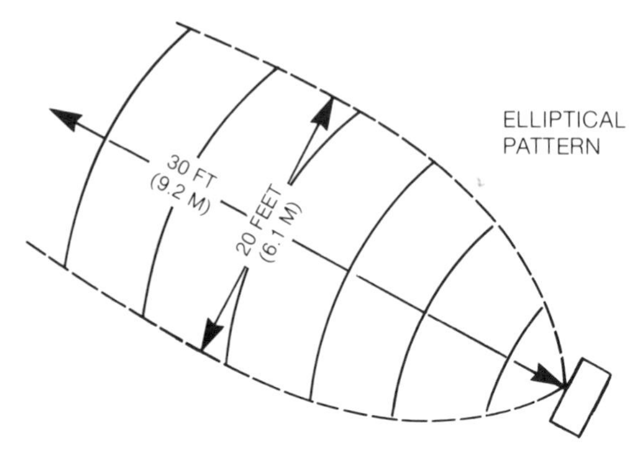

Figure 2-23. *Ultrasonic Energy Pattern*

Figure 2-24. *Picture Of A Microwave Space Detection System*
(Courtesy Radio Shack)

Space detectors are extremely difficult for intruders to defeat since there are no switches or wires to tip off the presence of an alarm system. In the early days of ultrasonic detection false alarms occurred very often, but the complex circuitry used in today's models has greatly reduced the problem. Still, by the nature of an ultrasonic system the following items may cause false alarms if they are very close to the unit: turbulent air from an airconditioning or heating unit, loud noises which contain ultrasonic energy (in rare cases a telephone bell may set the unit off), or vibration of the unit. With an adjustment, the sensitivity of most units can be reduced to avoid the false alarms and still give good intrusion detection. Because there is the possibility of false alarms many ultrasonic units use only localized alarms. The noise is sufficient to alert the occupant and to frighten off an intruder, but will not awaken an entire neighborhood.

Microwave Space Detectors

The fact that movement in a wave field produces a frequency shift in the reflected wave has already been mentioned. This same principle is used at extremely high frequencies – billions of cycles per second. These detectors are called radar or microwave depending on the frequencies used because they operate in the frequency range of the RADAR described. The coverage of microwave detectors is typically up to 40 or 50 feet – greater than that of the ultrasonic systems. Their envelope pattern of radiation is very similar to the ultrasonic pattern of *Figure 2-23*. Microwave radiation can penetrate walls which do not contain metal screening or metal surfaces, for example, brick walls, but it may be stopped by insulation with a metal foil backing. Microwave detectors operate between about 900 MHz (million cycles per second) and 22 GHz (billion cycles per second). The lower the frequency the greater is the penetrating power of the beam. Thus a 950 MHz unit can be expected to penetrate a brick wall while a 10.5 GHz unit probably would not. A typical unit is shown in *Figure 2-24*.

The amount of penetration obtained from a unit can be an advantage or a disadvantage, depending on how the unit is to be used. Often units which will not penetrate a brick wall are desirable, because the protected area can be contained inside a brick structure. The penetration power of all microwave detectors allows the unit to be placed out of sight from any intruder behind a cloth or paper. The penetrating power allows coverage of more than one room in a home or office in some cases. This can be a significant advantage from the standpoint of ease of installation and may save buying a second system.

Many of the early microwave detection systems used vacuum tube oscillators and did not have the reliability and maintenance free operation of the present solid-state systems, especially the systems that use a solid-state diode called a Gunn diode to generate the microwave energy. The Gunn diode systems operate at about 10 GHz. There are also some systems that use discrete solid-state oscillators that operate at 2.5 GHz and others at 915 MHz. However, very few of the lower frequency units are still manufactured.

Microwave detectors have significant advantages because they are not affected by air currents or noises. More care must be taken in locating the unit correctly since the protected area extends beyond barriers that seem solid. It also may extend to areas that one may not wish to control, such as a busy apartment hallway or the driveway next door. A good walk test is necessary to determine the actual area being protected. More care must also be taken in mounting the unit because vibration may cause false triggering and fluorescent lighting may call for some shielding.

Passive Infrared Systems (PIR)

"Hot body" infrared systems were mentioned previously. As strange as it may seem everything gives off infrared light energy. The hotter an object the more infrared energy emitted. The human body gives off infrared energy radiation of approximately 3×10^{13} Hz (this corresponds to about 90° F (32° C).

Detection systems are available that can "see" this energy pattern. Since the detection system does not transmit any energy pattern but only receives it from the bodies that radiate it, these systems are called passive.

Early Type

An early type of infrared system worked by placing a pyroelectric detector behind a diffraction grating. The diffraction grating consisted of a piece of material transparent to infrared light which had a very fine screen pattern etched on its surface. As infrared light entered the diffraction grating, it was bent and interfered with other infrared waves which had also entered the grating. This produced a pattern of a series of points of high and low energy. If an object moved even slightly, the interference pattern changed drastically and the current through the pyroelectric detector reflected the change. In other words, these moving shadows appearing as electrical pulses emitted from the detector were used to trigger an alarm.

Newer System

Figure 2-25 shows a new passive infrared system which uses several mirrors, each of which sees broken areas (fingers) of the room. The infrared energy is reflected from these mirrors through special lens made of semiconductor material (germanium) and onto the detector. Expensive modern units use more than one detector and use differential amplifiers to avoid false alarms. The effect is the same as with the diffraction pattern. If an intruder enters the pattern, the pattern is changed and this is detected by amplifiers and sent to a circuit which sounds the alarm.

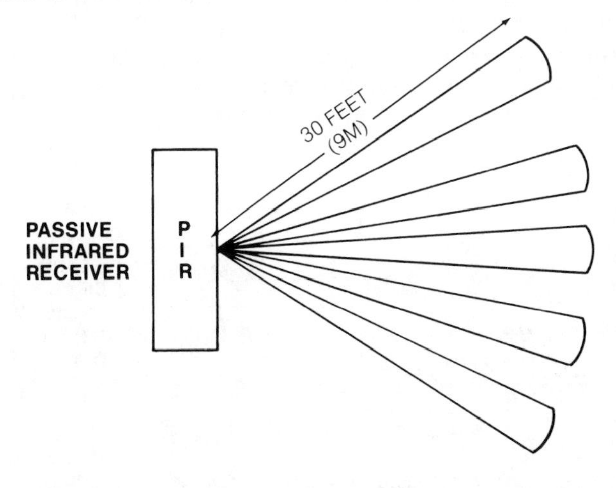

Figure 2-25. *Fingers Of A Passive Infrared System*

Passive infrared systems can generate false alarms when a change in the infrared energy pattern is not due to an intruder. For instance, direct sunlight entering through a window and shining on the detector might be detected as an intruder.

Passive infrared systems are fairly new to the industry, but they offer great potential. They require little power, are not subject to false alarming when properly placed, and potentially could be offered at a desirable price.

SEISMIC DETECTION

A variation of the seismic transducers that are used to detect and measure earthquake tremors can also be set up to detect the presence of a motor vehicle or the presence of an intruder. Seismic transducers contain a piezoelectric crystal (similar to that used in a microphone) encapsulated in a protective sheath. An electrical signal is generated when the crystal is stressed. As shown in *Figure 2-26*, the changes in pressure generated by humans or vehicles walking or driving on a surface is propagated to the piezoelectric crystal which detects the changes. These changes are then communicated by amplifiers and formed into a signal that sounds an alarm. With respect to a human intruder, typical coverage is a circular pattern having a 25 to 200 foot (7.6 – 61M) radius. If protection against motor vehicles is desired, circular areas having a radius of 1,500 feet (0.46KM) can be obtained. Soil conditions affect the amount of coverage since seismic waves have difficulty passing through frozen or extremely hard ground. Most units have a sensitivity adjustment which allows selecting the amount of area to be covered.

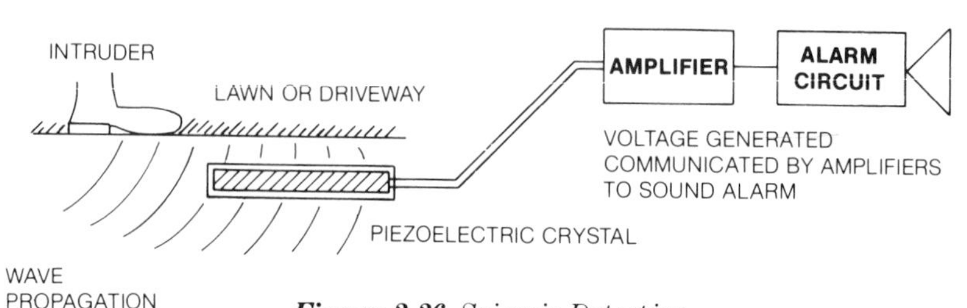

Figure 2-26. *Seismic Detection*

VIDEO SYSTEMS

Industrial security often controls entry into a location with video cameras placed at each entry *(Figure 2-27)*. A guard sees each person who attempts to enter and checks for proper identification before allowing them access by activating a remotely controlled lock. Compact cameras and TV-type displays also allow visual coverage of a large area. For unattended monitors, complex electronic units using photoelectric devices may be placed over the screen to detect movement (changes) in the light pattern and sound the alarm. There is nothing new in these specially adapted systems from standard TV principles except the equipment is especially designed for this application. A wide variety of different kinds of equipment for a system is available. Most systems are relatively expensive.

Figure 2-27. *Video System For Protected Access*

DETECTION BY A VARIATION IN CAPACITANCE

The human body has both an electrical resistance and capacitance. Circuits can be designed so that touching a portion of the circuitry changes the operation of the circuit, triggering an alarm. One such unit was attached to the inside door knob of a wooden door of Susan Clark's apartment in Chapter 1. Whenever the outside knob is touched the capacitance connected to the knob creates an electrical fluctuation and the alarm is sounded. Such units are relatively inexpensive. They will not work on metal doors unless the knobs are insulated, which is somewhat impractical. They also produce variable results because adequate direct hand or body contact is required to get the full body capacitance signal.

WHAT'S NEXT

Now that the various systems concepts are in hand, the discussion continues in the next chapter on how electronic detection systems are actually used.

Quiz for Chapter 2

1. Which of the following is not a basic function of an alarm system?
   a. Act.
   b. Communicate.
   c. Discuss.
   d. Detect.
2. Which of the following switches may be used in alarm circuits?
   a. Pressure sensitive switches.
   b. Heat sensitive switches.
   c. Mercury switches.
   d. Vibration switches.
   f. All of the above.
3. What is the primary advantage of a wireless perimeter system over a hardwired perimeter system?
   a. Digital coding.
   b. Initial cost.
   c. Ease of Installation.
   d. Signal modulation.
4. A Differential Amplifier is most commonly used in:
   a. A hardwired perimeter system.
   b. A wireless perimeter system.
   c. A sound detection system.
   d. An active infrared system.
5. A short-coming of a visible light system is:
   a. The system is unreliable.
   b. A thief could foil the system with a flashlight.
   c. False alarms are a problem.
   d. Everything emits infrared light.
6. A seismic detection system is similar to:
   a. A hardwired perimeter system.
   b. A microwave system.
   c. A sound system.
   d. An ultrasonic system.
7. The difference between passive and active infrared systems is:
   a. Passive systems react slower.
   b. Passive systems do not emit an infrared beam.
   c. Active systems do more work.
   d. Passive systems are less complicated.
8. Space protectors detect only objects which:
   a. Move.
   b. Emit infrared energy.
   c. Absorb energy.
   d. Shift gears.
9. Doppler shift principles are used in:
   a. Hardwired perimeter systems.
   b. Ultrasonic detectors.
   c. Microwave detectors.
   d. Infrared detectors.
   e. b and c above.
10. Video security systems are not often employed in home use because of:
    a. Poor picture resolution.
    b. Cost.
    c. Small area of coverage.
    d. Need of a cameraman.

(Answers in back of the book)

Personal Security for the Family

ABOUT THIS CHAPTER

Now that the need for basic security has been discussed and the general types of electronic detection systems available have been established, it's time to show some examples and to explore the reasons for selecting one system over another. Although several assumptions will be made in this chapter which may or may not fit a particular individual's circumstances, the analysis of the system chosen and the reasons it was selected will be of value in any case.

GROUND RULES

Let's assume that household and personal belongings, while valuable, are not the primary concern. Basically, the prime concern is to try to avoid a confrontation between an individual and an intruder. Everyone wants to feel that his home and family are secure from intrusion. If someone enters or attempts to enter, the occupants need to know. In addition, the intruder needs to be aware that his presence has been detected. Most intruders rely on working undetected, take away this advantage and they will flee.

Remember that there is no one best solution in the examples which follow. Several alternative plans may work equally well for about the same system investment.

CASE 1

Let's assume that *home* in this case is a rented one-bedroom apartment in a large suburban complex. The apartment is on the bottom floor, and there are neighbors on either side as well as above. The walls that separate the apartment's occupants from their neighbors are very thin, judging from the amount of noise which penetrates from the surrounding apartments.

The main entrance to the apartment is the front door which opens into a well travelled hallway. The door is secured with a medium-priced lock. Due to job transfers and other reasons, tenants stay in an apartment here on the average for less than a year, making key control difficult. Thus, tenants can't be sure that their key is the only one available. Also, it's well known that the landlord and certain maintenance personnel have pass keys.

The apartment has a sliding glass door that leads to a patio. The patio door serves as the only window for a good view of the outside and the shrubs and potted plants growing there. The occupant has a pet – a cat – which has free run of the home, not only the floor and bed, but it climbs on the furniture as well. The floor plan of the apartment is shown in *Figure 3-1.*

Now the question arises, what sort of security system would best fit the situation, with some concern for the pocketbook as well?

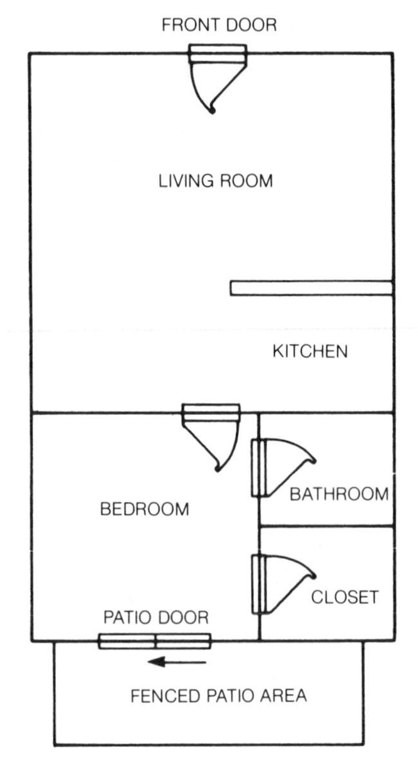

Figure 3-1. *Floor Plan Of A One Bedroom Apartment*

Space Detection?

The first system that comes to mind is some type of space detection (recall these units protect a space by beaming out an envelope or field of radiated energy). Since these systems are easy to install, a first choice might be to position the unit so the field could be beamed at the front door – the primary entrance into the apartment. Suppose a microwave system were chosen. As shown in *Figure 3-2*, this type space detector could present a problem because microwave systems generate a field that tends to penetrate through doors and walls. If such a system were directed toward the front door (first position) it might detect people merely walking up or down the hallway, providing many possibilities for false alarms. Let's look at a new position for the unit. The second possible position locates the unit so it beams across the living room. In this second position, the system would tend to penetrate the walls – possibly detecting movement in a neighbor's apartment and sounding an alarm. Both of these situations are undesirable because of the possibility of false alarms.

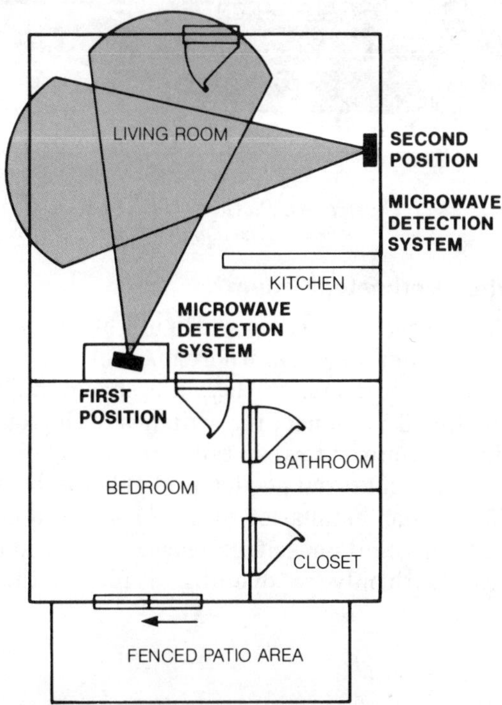

Figure 3-2. *Microwave Space Detection*

An ultrasonic space detector would overcome some of the problems of the microwave unit. It would not penetrate the door or walls and might be a better choice. *Figure 3-3* shows an ultrasonic space detector. Generally, systems of this type are available for purchase for under $100. But one unit in this case may not be totally satisfactory, because placing an ultrasonic system in the living room does not protect the patio door, a very vulnerable entrance. Both the microwave and sonic systems will detect the cat moving around. The cat would have to be confined in an area away from the detectors.

Figure 3-3. *Picture Of Ultrasonic Unit*
(Courtesy Radio Shack)

A Hardwired Perimeter System?

An alternative system which might be considered is a hardwired perimeter system where magnetic switches are placed on the front and patio doors. The layout is shown in *Figure 3-4*. Wires must be installed to connect the switches to the control box. Careful installation is required to keep the wires from being unsightly but it can be done. The material needed probably can be purchased for under $75. System installation probably will be a major part of the total cost. By doing it oneself, expense can be held to a minimum. For this case with only two openings, it may be the best choice.

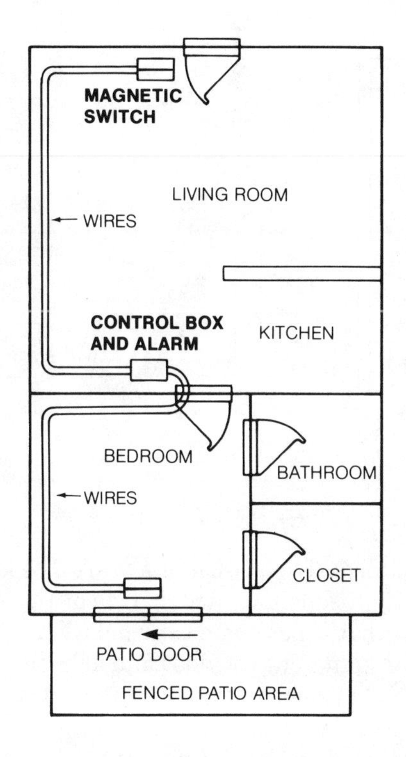

Figure 3-4. *Hardwired Perimeter System*

Simple alarms?

Simple alarms are another alternative for Case 1. As mentioned previously, they consist of a switch, a power supply and an alarm and are available in several different types. Where would they be placed? Certainly, one alarm would be placed at each door. The front door, a swinging door, could be secured in several ways. A first choice might be a doorstop alarm that would be placed against the front door whenever protection is desired. The doorstop alarm *(Figure 3-5)* sounds whenever the door is pushed into the spring-loaded switch. This alarm has the added advantage of effectively blocking the opening of the door.

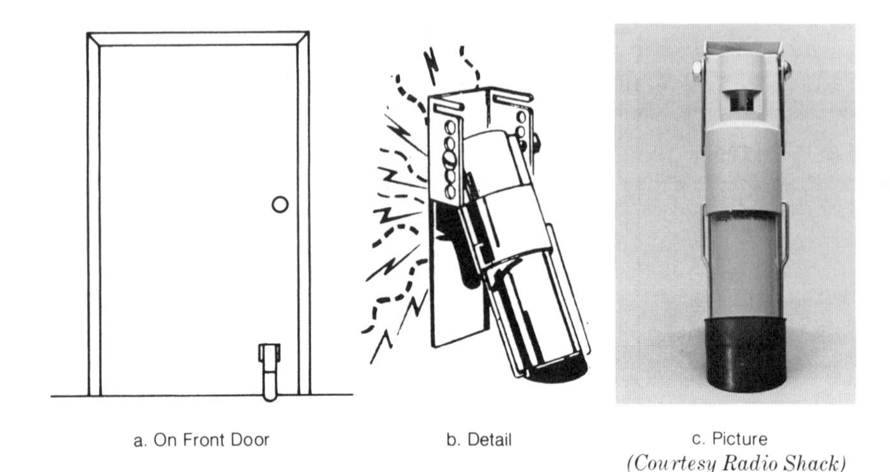

a. On Front Door b. Detail c. Picture (*Courtesy Radio Shack*)

Figure 3-5. *Doorstop Alarm*

Another alternative (shown in *Figure 3-6*) is a small unit which uses a vibration-sensitive switch. One of the advantages to this type of unit is that it has some time delays built into the unit. A 20-second delay for going out the door and an 8-second delay for coming in the door. This permits the unit to be armed and to detect intrusion through the door when the occupant is not present. This alarm could also be used on the sliding patio door, whereas the doorstop unit could only be used on the swinging type door.

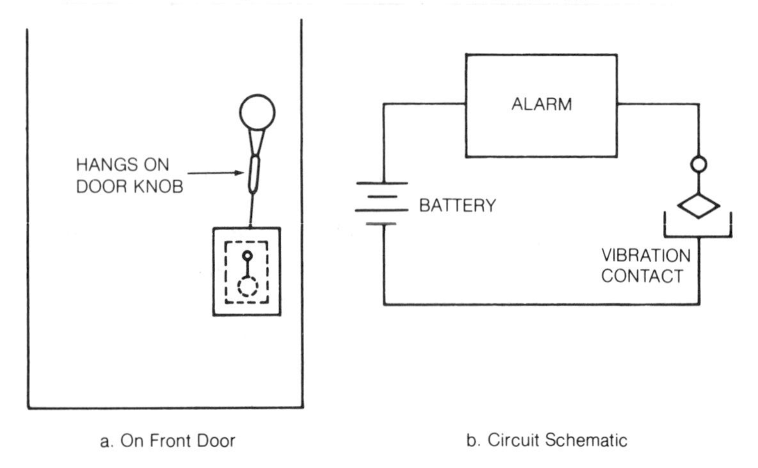

a. On Front Door b. Circuit Schematic

Figure 3-6. *Vibration Sensitive Alarm*

A third choice might be a trap switch alarm for the patio door. This is shown in *Figure 3-7*. In this normally-open switch type there is an insulator between the switch contacts keeping the circuit open. A string or some sort of actuator is attached to the insulator and to the patio door so that, if the patio door is opened, the actuator will pull the insulator from between the contacts sounding the alarm. A pet – the cat in this case – would not be a problem with any of these simple alarm systems, and they have another advantage – they can be purchased for a minimum cost, most under $50. Installation is extremely easy, although the alarms must be put in place each time they are used – probably as the doors are locked.

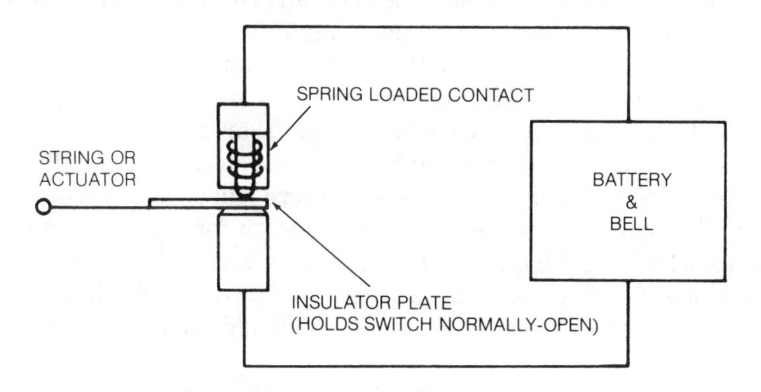

Figure 3-7. *Simple Alarm System With Trap Switch*

Typically, simple alarms placed on doors emit a localized alarm which stops sounding as soon as the violation stops. A persistent intruder could find the unit and smash it before the neighbors became sufficiently alarmed. With this situation in mind, it might be desirable to install a back-up system to continue sounding the alarm. A simple solution is to have panic buttons at strategic locations such as near the bed or near a favorite chair. The panic button (switch) is wired to a very loud alarm which sounds for several minutes after the button is pressed. The neighbors would certainly be alerted to the trouble when this second alarm sounded. A button alarm and wire for the panic alarm system also would probably cost in the neighborhood of $50.

In conclusion, if home is a small apartment with a pet, close neighbors, and protection is desired principally when at home, simple alarms may be the most sensible (and economical) choice.

CASE 2

Let's suppose that there has recently been an addition to the family and more room is required. As a result, *home* now is a leased, small 3-bedroom, 2-bath home in a residential area of a medium-sized suburban town. After renting the home, the family discovers that they are in an area of town that is experiencing an outbreak of daytime robberies. Only last week a house less than a mile away from home had a break-in. The intruders, thought to be members of a local gang, literally tore the place apart. They emptied every drawer in the house, poured soda pop into the piano, smashed the television set, and emptied the contents of the refrigerator on the kichen floor. A call to the local police not only confirmed the details of the robbery but revealed this alarming bit of information – daytime suburban robberies with associated vandalism are the most rapidly increasing type of crime.

As more and more wives work, suburban areas are increasingly deserted during daylight hours. In Case 2, the wife has quit her job to take care of the baby, so there will always be someone home. This, however, causes additional concern for the husband for he fears what might happen if the gang discovers that his wife is home alone. The first concern was to install sturdier locks, but the landlord refused because of the expense. He told the family they could take the added precautions but at their own expense. They investigated, found hollow, unsturdy doors, and decided to look for an electronic detection system instead. The home floor plan is shown in *Figure 3-8*.

Simple Alarms?

As in Case 1, maybe simple alarms could do the job? A quick count of the openings into the home shows ten windows, the front and patio doors and the garage door. The number of simple alarms required and the idea of setting up and maintaining so many of them makes the use of simple alarms out of the question for overall perimeter protection.

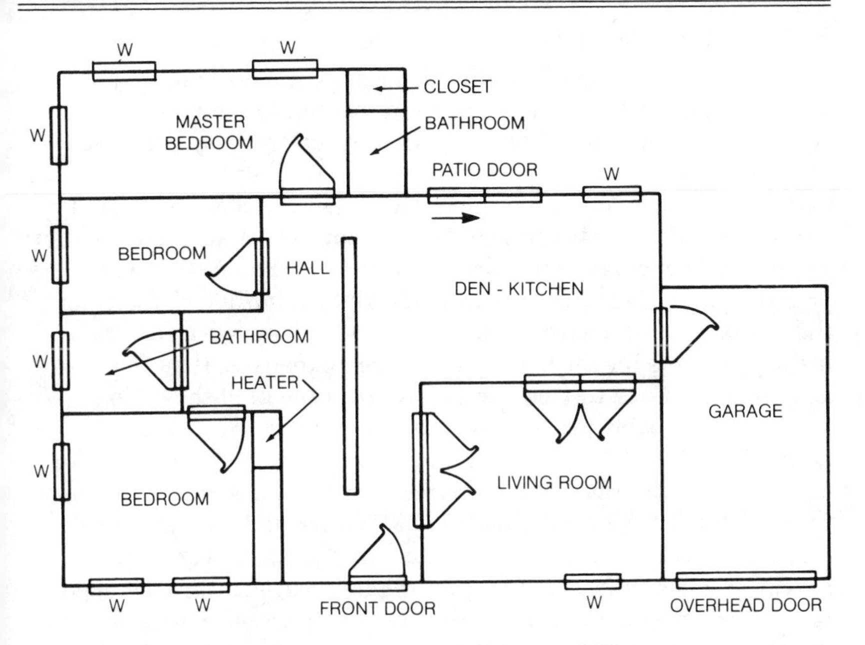

Figure 3-8. *Floorplan Of The Rented Home*

Space Detection?

How about a space detection system? This certainly would be fine for areas of the home at night, but with the wife at home using the house during the day it would be very restrictive on the use of the home. The detection systems would constantly have to be turned on and off, or else traffic would have to be restricted from the area covered by the space detection system.

A Hardwired Perimeter System?

For this case, it seems that a good system would be a hardwired unit which protects all the windows and doors, and allows free access in the house. But recall that a concealed installation of a system in this existing home would be difficult calling for extensive installation with its associated expense. Bear in mind the home is rented. The family certainly will not stay here indefinitely. If a hardwired system is installed, a considerable investment in both time and money will be left behind when the family leaves.

A Wireless Perimeter System?

A system that offers the advantages of a perimeter system but eliminates much of the permanent installation is a wireless perimeter system. In one type of wireless perimeter system, each transmitter plugs into an electrical outlet. Another type has a battery for power and can be placed anywhere. If electrical outlets are at a premium, or if transmitters are not to be restricted to outlet locations, the battery transmitters for the wireless system will probably be the best solution. This system will include virtually all the advantages of a hardwired system and will be much easier to install. Systems for the home will be more expensive than for the apartment of Case 1. Good wireless units of the kind shown in *Figure 3-9* probably are available for around $200. Such units normally include two transmitters with switches, a wireless smoke detector, a panic transmitter, the receiver and the alarm. Additional transmitters are available for around $30 each and good switches cost from three to five dollars apiece. Many of the newer units provide digital coding of the radio frequency signal – giving good false alarm protection. Many units provide an on-off switch for each transmitter. This is handy if the doors or windows in a particular area are to be used while the rest of the house is under protection.

If the family in Case 2 decides on a wireless system, they will be able to take the system with them when they leave and install the same system in their next home. Let's assume that was the choice and look further into how such a system is installed.

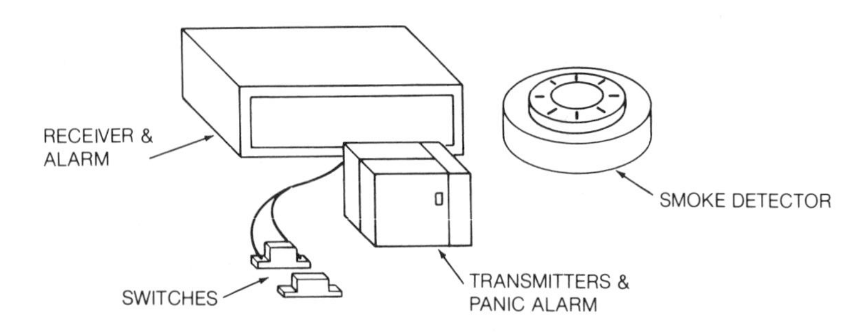

Figure 3-9. *A Basic Wireless Kit*

Installing The System

The switches provided with a wireless perimeter detection system are probably of the normally-closed magnetic variety as shown in *Figure 3-10*. They are encased in plastic and consist of a metallic arm which closes the contacts when a magnet is near. When the magnet is moved away from the switch (a door or window is opened) it no longer holds the contacts together and the spring pulls them apart. These switches are often filled with an inert gas which keeps the contacts from coroding for many years. Plunger type switches do not have this advantage and after several years may cause intermittent operation which could be difficult to find.

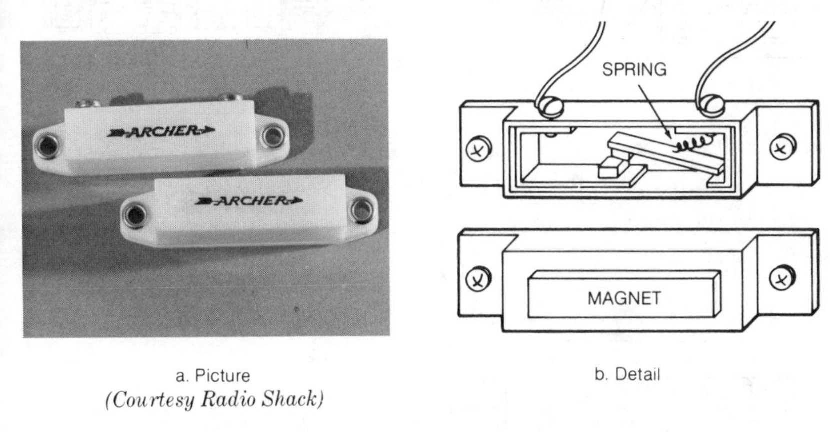

a. Picture
(Courtesy Radio Shack)

b. Detail

Figure 3-10. *A Typical Normally-Closed Switch*

If every opening in the home had to have its own transmitter, the cost of a wireless system would be prohibitive. However, several openings can be wired to one transmitter. The master bedroom in the home has three windows which need protection; the switches can be wired to one transmitter as shown in *Figure 3-11*. The other rooms could be wired in a similar manner as shown in *Figure 3-12*. This complete system would not have to be installed at one time, and certain areas might be left unsecured as unlikely points of entry. Maybe the front bedroom windows face a busy street and are difficult to reach, or the bathroom and bedroom windows are next to a streetlight. If the main garage door had a garage door opener, the door to the house from the garage might not need protection. In Case 2, if all perimeter openings are protected, the approximate cost of the wireless system is shown in *Table 3-1*.

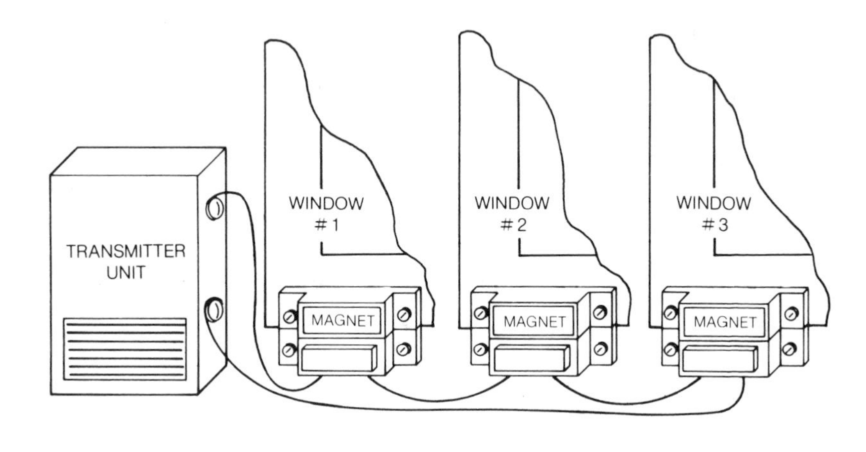

Figure 3-11. *Wiring Connections For The Master Bedroom*

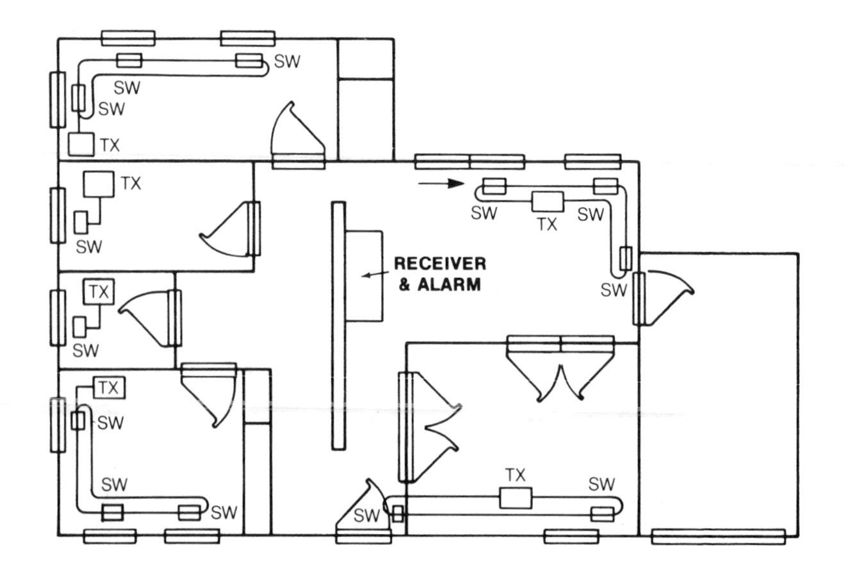

Figure 3-12. *Placement Of Switches And Transmitters*

	Total Req	With Kit	Must Buy	Total Cost
Basic Kit (including 2 Transmitters and 2 switches)	1			$200
Transmitters used (Buy 4 at $30)	6	2	4	$120
Switches used (Buy 11 at $3)	13	2	11	$ 33
Wire (Approximately)	1 Roll		1 Roll	$ 10
Approximate Total Cost				$363

Table 3-1. *Approximate Cost of Wireless Perimeter System*

The Smoke Detector

Included in the wireless perimeter system for Case 2 was a smoke detector. It is one of the most important security alarms developed in the past few years. It is an electronic detection system, but of a type not yet discussed. In American homes 700,000 fires strike each year at a rate of over one per minute. A rather startling fact is that of the people who die in fires, 90% die in their own home. They are not consumed in flames as you might imagine, their lungs are filled with poisonous gases. About 75% of household fires start as slow smoldering fires – the result of spontaneous combustion of oily or dirty rags, or the result of heat produced from a faculty electrical connection or furnace, or the careless use of space heaters or because of smoking in bed. The fumes spread throughout the house long before the flames. People asleep in the house will likely be overcome by fumes before they are awakened if the home is not equipped with one or more smoke detectors.

Technological breakthroughs in the last few years have made smoke detectors very efficient. There are two basic types of smoke detectors in common use: ionization and photoelectric. Both types provide good detection at low cost (under $20 in some cases). Ionization detectors normally respond somewhat faster to flaming fires which produce little visible smoke and photoelectric detectors respond quicker to smoldering fires. For most home use this gives the photoelectric detector a slight edge – since 75% of home fires are of the smoldering type. On the other hand, photoelectric detectors tend to use more current and may require batteries more often (or an ac connection). Certain brands of ionization and photoelectric detectors may be superior to others, but all those which carry the approval of a nationally recognized testing service such as Underwriters Laboratories or Factory Mutual should be satisfactory.

Some detectors have a built-in test capability that is nice to have. A simple test for a smoke detector is possible by holding a match or cigarette under it to determine if it is working. This does not verify completely that it is working properly, since the amount of smoke supplied with a match or cigarette is many times that necessary to set off the detector when it is working correctly. Nevertheless, this simple test provides some assurance that the detector is working. For the smoke detectors that are battery operated, many have a small push-button switch, which when closed sounds the alarm if everything is in order. If no alarm sounds, the first step would be to replace the battery. *Figure 3-13* shows an electrical detection system that uses an ionization chamber as the detector for smoke particles.

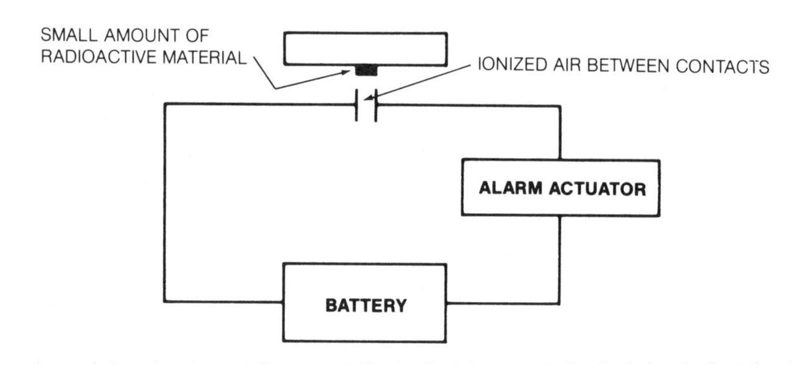

Figure 3-13. *Ionization Type Smoke Detector*

How Do Smoke Detectors Work?

Ionization detectors use a small amount of radioactive material to ionize the air adjacent to them. As long as the air is ionized it conducts electricity and a current flows through the alarm actuator and the battery. As long as this current flows the alarm does not sound. If smoke enters the chamber it reduces the number of ions (smoke particles attach themselves to ions) which are free to carry the current. This reduction in current is sensed by the alarm actuator and the alarm sounds.

Photoelectric detectors use a small light and a device that changes resistance when light shines on it. Normally the photocell is placed in the end of a hollow cylinder. A small light on the photoelectric detector shines toward the photocell, but the path is blocked by a portion of the cylinder *(Figure 3-14)*. When smoke particles enter the cylinder they scatter and reflect the light onto the photocell causing a current change which is amplified and used to sound the alarm.

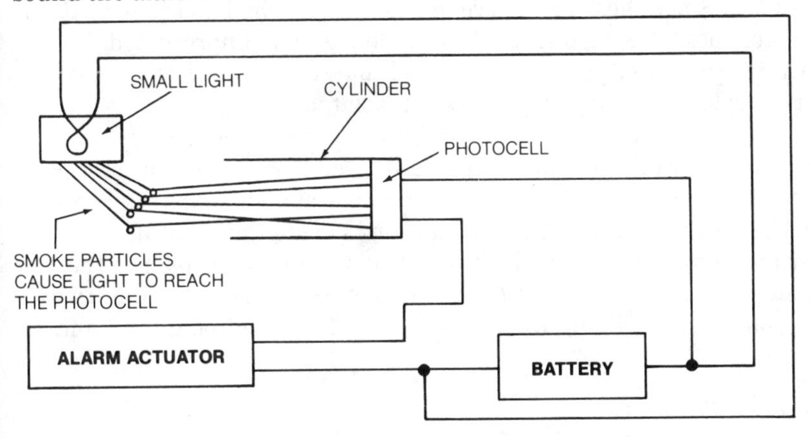

Figure 3-14. *Photoelectric Smoke Detector*

Installation and Maintenance

Most smoke detectors are usually less than ten inches in diameter and very light in weight. Most are powered by batteries making the unit independent of home electrical needs. A smoke detector should be mounted on the ceiling of the highest point in your home – upstairs if you live in a two-story home. In many cases the ceiling outside your bedroom or the hallway leading to all bedrooms is a desirable location. A smoke detector should not be installed in the kitchen or in a room with a fireplace since smoke is often present in these areas. Generally the base of the smoke detector is mounted to the ceiling with screws and then a cover snaps over the base. All that need be done to change batteries is snap off the cover and install new batteries. Most smoke alarms give a warning beep when the battery current becomes low. Batteries should be replaced when the beeping occurs. Some smoke detectors use very expensive batteries that are difficult to find. Usually it is better to purchase one that has a standard battery.

Let's return now to the application of additional types of electronic detection systems for security purposes.

CASE 3

In the example of Case 3, a husband and wife, or either one, have ended up with a larger house than is needed. The kids are grown or there has been a death or separation in the family – which leaves the occupant with a home in which several rooms normally are not used. Since these rooms are not closely watched and they contain valuable possessions, one hesitates to leave them unprotected. Not only are they easy prey but they tend to be an easy access for the intruder because they are somewhat isolated.

A large existing house such as this would require quite a large number of components if a hardwired perimeter system were used; and it would be very expensive to install in an existing home such as this. Therefore, the assumption is made that a total perimeter system is of no interest at this time. A floorplan of the home is shown in *Figure 3-15*. The rooms which are not being used regularly are the living room and the two front bedrooms, #1 and #2.

Space Detection?

One solution to providing protection would be to use a partial perimeter system just in the front of the house, and this could be a good solution except for the installation and disruption it causes. A more acceptable approach would be to use a space detection system. Let's see how this might be done.

Protection is desired for three rooms. If an ultrasonic unit is used, because ultrasonic would not penetrate walls, one would have to be used in each room. With a high-power microwave system, all three rooms can be protected with the same system. Industrial type microwave detectors have a range up to 100 feet through air. When a wall is penetrated the range is reduced. However, it should be sufficient to cover the 40 feet required for this case. The actual area covered can be determined by placing the system in the test mode and walking around the area to see if the alarm sounds. The best position for the system in the home seems to be with the unit located as shown in *Figure 3-16*.

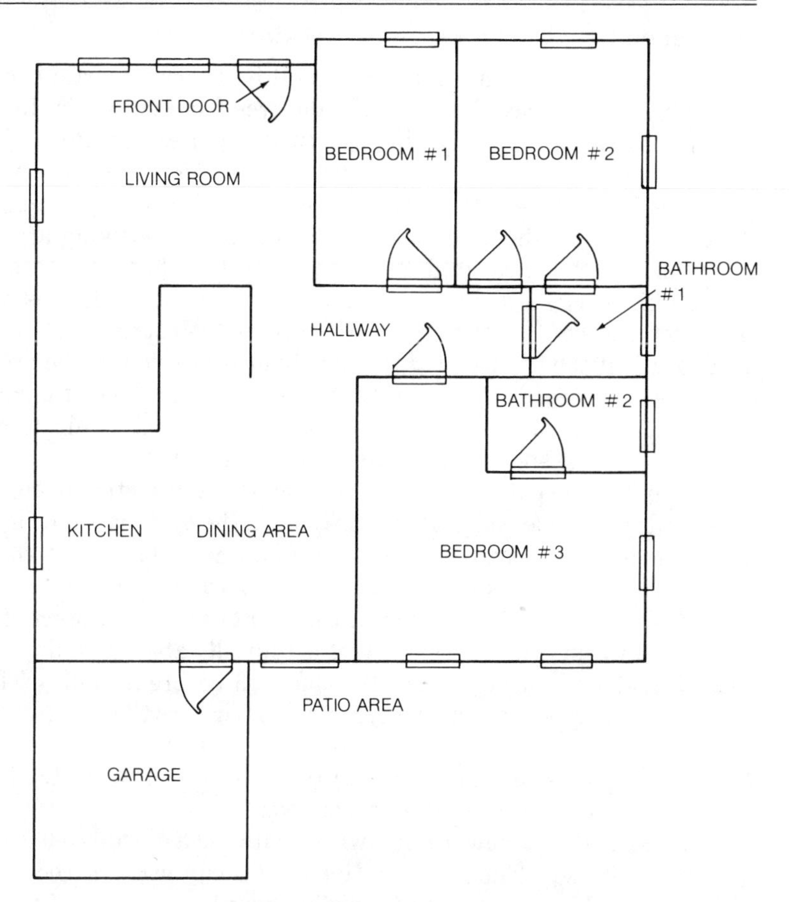

Figure 3-15. *Floor Plan*

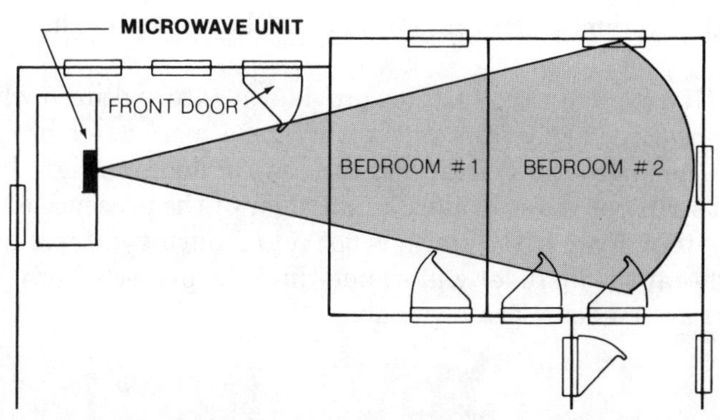

Figure 3-16. *Location Of A Microwave Unit*

Installation

Recall that with a microwave unit it is very important to have a mounting place that is stable and free from vibration. In addition, it can be hidden behind a cloth or paper screen. Power is applied by plugging the cord into an ac outlet and interconnecting the wires for the siren. As stated earlier, the unit can be placed in the test mode and the detection pattern verified by walking around the rooms which are covered. When in the test mode many systems have a light which indicates when the alarm is sounded. In this way the exact areas of coverage can be determined. Most units allow the user to adjust the intensity of the field in order to control the area of coverage. Therefore, with the intensity control and the test mode, the specific pattern required can be obtained. The only limitation would be if the system did not have enough range at full intensity.

After the pattern is adjusted correctly the system need only be armed to provide full protection. To arm, the system is put in the active mode. There is about a thirty second time delay before the unit is activated. This gives the user time to leave the protected area. A similar time delay occurs if the user enters the protected area when the system is armed. This is typically about twenty seconds and allows time to turn the alarm off before it sounds. The same time delay occurs when any intruder triggers the alarm. However, since the intruder does not know that the system has done the detecting it is very unlikely that the system will be located in time to turn it off before the alarm sounds.

Some of the newer microwave units use a digital code to disarm the system. This was mentioned in Chapter 2. The code is either entered by switches or from a keyboard on the face of the unit. The keyboard makes it virtually impossible for an intruder to disarm the system. Other systems may use key locks for this purpose.

Many microwave systems provide an external siren which should be located away from the unit. When a space detector is used there are no magnetic switches on windows or doors, or light beams, or metal strips, or wires to alert an intruder of the presence of an alarm system. Even if the alarm is not set off during entry, chances are good that the intruder will wander into the protected area as attempts are made to locate valuables.

CASE 4

One always dreams of building a new home. In Case 4, that dream is about to be realized. The proud owners have chosen a lot in a suburban area where they feel intrusion is not a threat. But to be safe, they have decided to install a protection system. They are wise because they are going to save themselves hundreds of dollars by planning the system and running the interconnecting wires while the house is under construction. Total system costs may also be less because the building contractor is installing the system.

It has been mentioned before that one of the primary drawbacks of a hardwired perimeter system is the installation of the wires. If a home is being built or remodeled, this problem no longer exists. Even if an electronic detection system is not installed immediately, it costs very little to have the interconnecting wires run in place during construction. The system can be installed later. Here are some guidelines for installing the wires *(Figure 3-17):*

1) Two wires should be installed to each perimeter opening.
2) Extra wires should be installed to well travelled locations for use with pressure sensitive or photo detectors.
3) Wires should be installed to furnace or water heater areas to accommodate heat sensitive switches.
4) Wires should be installed for locating smoke detectors.
5) Extra wires should be installed at strategic locations for panic buttons.
6) Wires should be installed for exterior lighting that may later be connected to an alarm system.

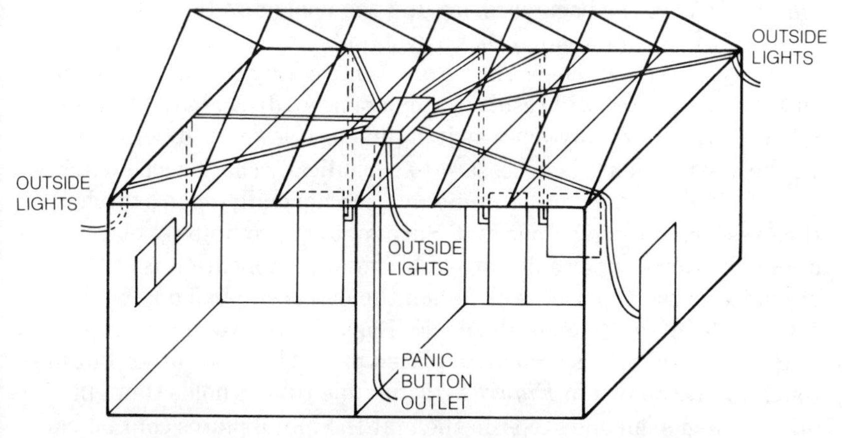

Figure 3-17. *Place Wires In A Home Under Construction*

During home construction is a good time to add some extra features. For instance, it's nice to have on-off switches by windows or doors that are more likely to be used while the alarm is on. As shown in *Figure 3-18*, the toggle switch shorts across the detector so that current can flow even when the magnetic switch on the window is activated due to opening the window.

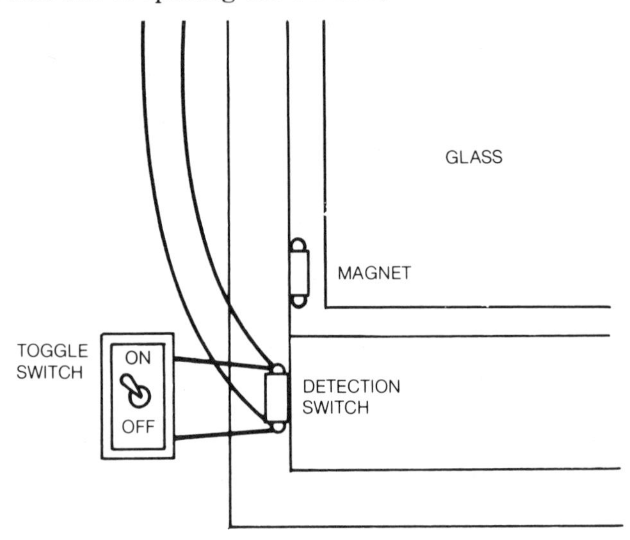

Figure 3-18. *A Switch To Disarm One Window*

The patio door might be a good place for another switch, so it can be used while the system protects the rest of the house when the family is in the back yard, around the pool or on the patio.

It is recommended that a system be installed that has normally-closed switches. They draw some current at all times and an alarm condition will result if they open. As discussed in Chapter 2, a defective normally-open switch (unless it is shorted) will not set off the alarm, thus it is necessary to periodically check such switches.

Perhaps the most commonly used normally-open switch is the pressure sensitive floor mat. Such mats are commonly placed under the carpet near a doorway. A common application is at the entrance to the supermarket. When the pad is stepped on, it activates controls to open the doors. Basically the switch consists of two thin metal plates which are placed parallel to each other but not touching. As shown in *Figure 3-19*, flexible rubber holds them in place. When someone steps on the mat the metal plates contact each other and the alarm sounds.

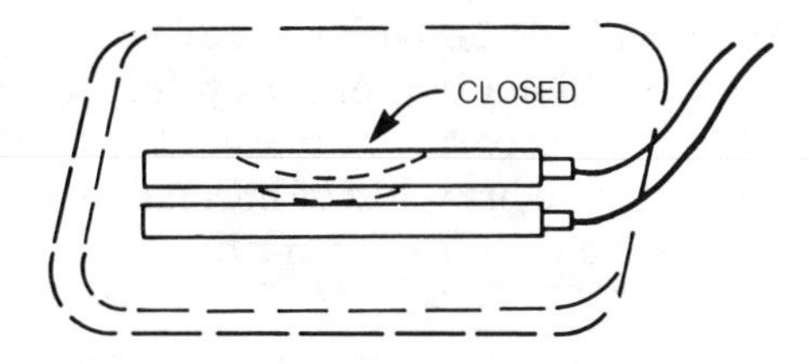

Figure 3-19. *A Pressure Sensitive Switch*

Temperature sensitive switches are used in a simple alarm circuit to sound an alarm when there is fire in areas such as hot water heater closets or furnace areas. They are normally set to close at about 130°F (54.5°C). They work much like the automatic choke on a carburetor in an automobile. A coil spring is constructed from a metal strip which is actually two metal strips welded together. One side of the strip expands more than the other when heated. *Figure 3-20* shows a temperature sensitive switch where the inside metal expands more than the outside when the coil is heated. This causes the coil to unwind forcing the contacts together.

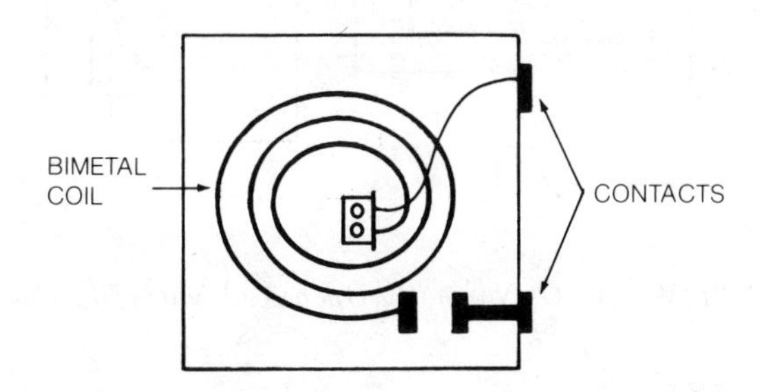

Figure 3-20. *A Heat Sensitive Switch*

Wiring the Switches to the Control Box

The control box supplied with a hardwired perimeter system should provide terminals for both normally-open and normally-closed switches. Recall in Chapter 2, a system was discussed that handled both types in separate branches. The wiring scheme which is used to connect these branches of normally-open and normally-closed switches is shown in *Figure 3-21*.

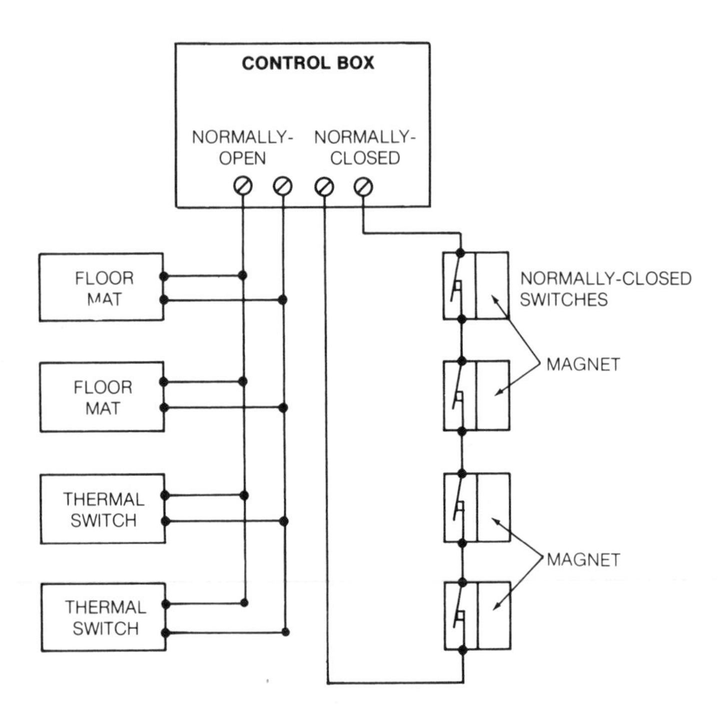

Figure 3-21. *Wiring Of Normally-Open And Normally-Closed Switches*

WHAT'S NEXT

In the examples discussed for the homes chosen, the main object was to avoid confrontation between the occupant and the intruder and to take the element of surprise away from the intruder and give the occupant warning needed to take action. That is, the homes were occupied. The next chapter will center on protection of property when the owner is not present.

Quiz for Chapter 3

1. Chapter 3 suggests systems which will protect primarily when:
   a. No one is present.
   b. Valuable merchandise is to be protected.
   c. Someone is present.
   d. One fears vandalism.
2. Systems that protect exterior openings of a home are called ________ systems.
   a. Perimeter.
   b. Space.
   c. Microwave.
   d. Ultrasonic.
   e. All of above.
3. Hardwired systems are systems that are installed:
   a. To allow free access in the home.
   b. When soft wire is not available.
   c. With wires to interconnect components.
   d. a and c above.
4. A primary advantage of a doorstop alarm is:
   a. It can be set up from the outside.
   b. It looks impressive.
   c. It is difficult to install.
   d. It impedes the opening of the door.
5. Primary advantages of a vibration alarm are:
   a. It is easy to install.
   b. It is portable.
   c. It can be used on any door.
   d. All of the above.
6. A trap switch is:
   a. Easy to use.
   b. Economical to purchase.
   c. Very difficult to foil.
   d. Very reliable.
   e. a, b and d above.
7. A ________ system allows pets to roam free in a home.
   a. Microwave.
   b. Ultrasonic.
   c. Perimeter.
   d. Space.
   e. None of above.
8. In Space Protectors "space" refers to:
   a. The fact that space age technology is used.
   b. The unit itself occupies a small amount of space.
   c. The unit was developed as part of the space program.
   d. A space or volume is protected by the unit.
9. In protecting a small area a microwave space system may cause problems because:
   a. The unit is difficult to find.
   b. The noise may be a problem.
   c. The protective beam may penetrate into adjoining areas.
   d. The coverage is very thorough.

10. An electronic detection system which installs by plugging units into electrical outlets is a type of:
a. Hardwired system.
b. Wireless perimeter system.
c. Infrared system.
d. Passive Infrared system.
e. None of above.

11. The primary disadvantage of a wireless system is:
a. System appearance.
b. Locating the unit.
c. Installation.
d. Cost.

12. Of the people who die in fires, what percent die in their homes?
a. 40%.
b. 50%.
c. 60%.
d. 90%.

13. An ionization smoke detector uses a small amount of:
a. Gallium.
b. Photoconductor.
c. Radioactive material.
d. Silicon.

14. A photoelectric smoke detector relies on ________ to reflect the light onto the photocell.
a. A mirror.
b. Smoke particles.
c. A cylinder.
d. An alarm actuator.

15. A smoke detector should not be installed in:
a. A living room.
b. A hall.
c. A kitchen.
d. A bedroom.

16. The majority of the protective switches in a perimeter system should be:
a. Normally – closed.
b. Normally – open.
c. Pressure sensitive.
d. Actuated by capacitance.
e. A toggle type.

17. Normally – closed magnetic switches are filled with an inert gas to:
a. Preserve the electromagnet.
b. Ionize electrons.
c. Keep the switches from coroding.
d. Help the contacts float.

18. A coil spring heat-sensitive switch is based on the fact that:
a. Rubber shrinks as it is cooled.
b. Electrical resistance increases with temperature.
c. Pressure times volume equals temperature.
d. Metals expand at different rates as temperature increases.

19. A pressure sensitive switch placed in a rubber mat is a ________ switch.
a. Normally-open.
b. Normally-closed.
c. Toggle.
d. Magnetic.

20. The best time to install a hardwired perimeter system is:
a. After a robbery.
b. During the summer.
c. During home construction.
d. When remodeling.
e. c and d above.

(Answers in back of the book)

Home and Business Property Security

ABOUT THIS CHAPTER

So far the applications of detection systems which protect property while the occupant is on the premises have been discussed. This chapter explores the problem of protecting a home or business while no one is present. Intruders normally strike a business at night, after working hours, or on a weekend. For homes it seems to be different. The opportune time for the intruder is anytime the home is not occupied, whether day or night. Why a particular home or business is selected by the intruder(s) many times remains a mystery.

The common sense procedures to follow when homes are left idle for several days or weeks have already been covered. Many of these same steps apply to a business as well. In particular, the police department should be notified when a business is to be shut down for a period of time. By so doing, they can pay particular attention to it while they patrol. Simulating an occupants presence by operating lights or music or air conditioners or machinery at periodic times can be a deterrent but a detection system probably will be more effective.

WHAT SYSTEM SHOULD BE INSTALLED?

When an alarm system is installed, the first question that must be asked is "Who will respond when the detection occurs?" Because no one is present, the system, besides sounding a local alarm, must now communicate to the outside to get help. Will it be the police, someone nearby, a neighbor, or a central security location? Whomever it is, what is important is that the responder follows through quickly.

A wide variety of possible systems exist. They have a variety of outputs. Some turn on lights, others sound outside bells or sirens, others activate remote telephone dialers which place a call for help. Hopefully, in some systems the alarm would scare away the intruder. In other systems the intruder would be unaware that he has been detected. The type of detection system chosen will depend on how isolated the property is and how valuable the items are that are contained in the home or business. Let's look at some specific cases.

CASE 1

The first example concerns the college residence of a student who lives in a large boarding house. *Home* is one room in a large and busy house *(Figure 4-1)*. The most valued possession is an expensive stereo system which needs to be protected. Traffic through the house is almost uncontrollable. It's nearly impossible to tell who is a visitor, who is a friend or who is a neighbor. Obviously, installing new locks on the doors and windows is the first priority. But these may be of little value, door and window frames are not at all solid and screws can be torn from the frames very easily. Is there a system that will protect this home?

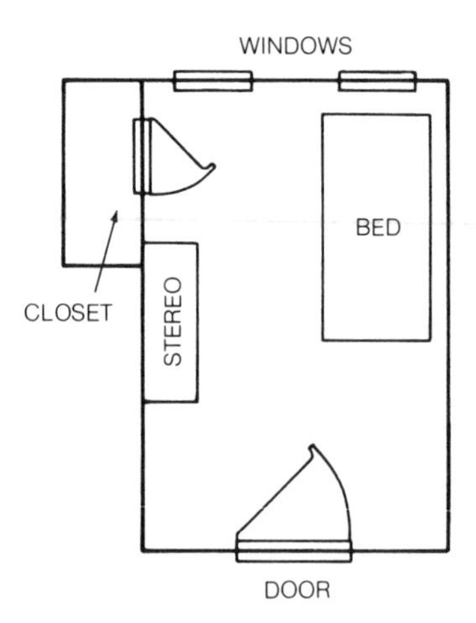

Figure 4-1. *Floorplan Of College Student's Room*

Simple Alarms or Perimeter Systems?

Simple alarms can be placed on the door and windows. Three detectors would be necessary. This certainly would be satisfactory but it calls for lots of persistence because the alarms must be placed in position each time the occupant leaves and college rooms are busy places.

A perimeter system also is a solution. However, because the room is only going to be home for another year, the time, effort and cost may be more than the occupant can muster at this time. On the other hand, if unused surplus parts can be obtained from particular university departments and time is available, this might be a viable inexpensive approach.

A Space System

The easiest system to install and use in this case is a space system. An ultrasonic system is the logical choice because a microwave unit would penetrate walls and doors and is likely to give many false alarms due to the close proximity of neighbors in adjacent rooms and in the halls. One ultrasonic unit, positioned as shown in *Figure 4-2*, would cover nearly the entire room. The radiation pattern would be such that an intruder would surely disturb the pattern as he moved to any part of the room. The local alarm which comes with the unit would be all that is needed in this case since a nearby neighbor will almost always be on hand in such a busy place. One note of caution: if there is a window air conditioner the sensitivity of the unit will have to be reduced since the turbulent air could cause an ultrasonic unit to sound a false alarm.

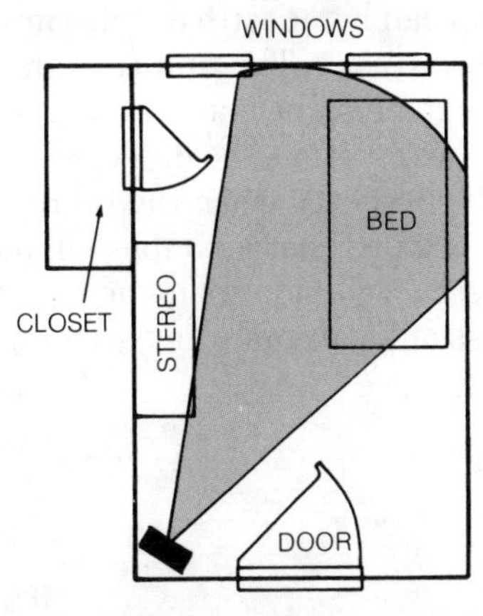

Figure 4-2. *Ultrasonic Protection For The Room*

Installing and Using an Ultrasonic Detector

Most ultrasonic units are equipped with both an exit and entry delay. When the system is turned on there is about a twenty second exit delay before the unit will operate. This gives the occupant time to leave the room after arming the system. Upon return, there will be a fifteen second entry delay before the alarm sounds. This allows the occupant to switch off the unit after entering to prevent the alarm from sounding.

For installation, place the unit on a sturdy foundation, plug it in, and aim it toward the area that is to be covered. The initial tests should be conducted with the entry delay disconnected (see the instructions shipped with the unit). Turn the unit on, wait until the exit delay has activated the unit then proceed to move around the room noting the positions that cause the alarm to sound. Many units have a special circuit condition called "TEST" for this purpose. Switching to "TEST" disconnects the alarm but substitutes a light that signals when the alarm is activated. The total field of coverage can be determined and adjusted by testing the pattern over the whole room. If the range needs adjustment to extend or retract the pattern, or if the position of the unit must change, further tours around the room will result in the pattern desired. Reconnect the entry delay when all testing is complete and the unit is ready to use.

CASE 2

In many applications, there is not just one detection system that satisfies the need. A number of systems can do the job. Case 2 is such an example, and many homes match the conditions described.

Case 2 is a suburban home with 3 bedrooms and 2 baths. The floor plan is shown in *Figure 4-3*. The family in this home have neighbors nearby who can hear the alarm and have agreed to take action when it sounds. Let's look at the detection systems that might be chosen to protect this property when the family is not present.

As mentioned in a previous example, simple alarms are probably out of the question for a large home because of the number of units required, unless only very minimal coverage is desired.

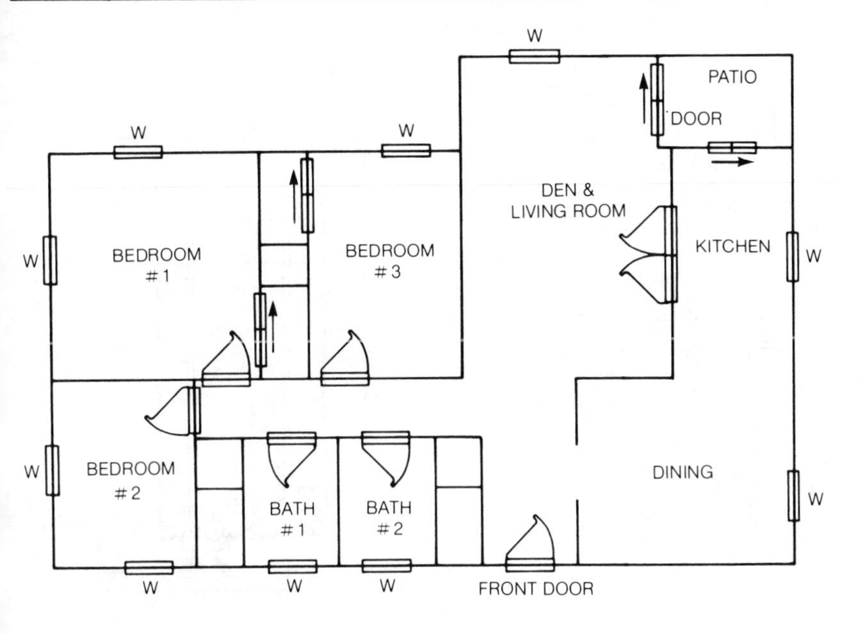

Figure 4-3. *Floorplan Of A Suburban Home*

A Perimeter System

A hardwired perimeter system is totally satisfactory and would give good protection. The common objection, as before, is that it is somewhat difficult to install. If a hardwired system is chosen, a few further comments should be made with regard to securing windows using the hardwire switches. For a home where no one is present, an intruder may not be afraid to completely smash out a window and crawl right through the frame without raising the window. In this case, if the window is equipped with a magnetic switch, the alarm would not sound because the magnet was not moved to trip the switch. There are several solutions to this problem. One is to use a backup space detector in a hallway to catch the intruder after entry. Another is to "tape" the window. As discussed for the commercial store front doors and windows in Chapter 2 *(Figure 2-8)*, the tape looks like silver masking tape. It is a conductor of electricity but is brittle and breaks whenever the glass under it is broken. The tape is installed by carefully marking a pattern on the glass with a grease pencil and then sticking the tape alongside the mark. Usually the tape is placed about 3″ from the edge of the glass around the perimeter of the window as shown in *Figure 4-4*.

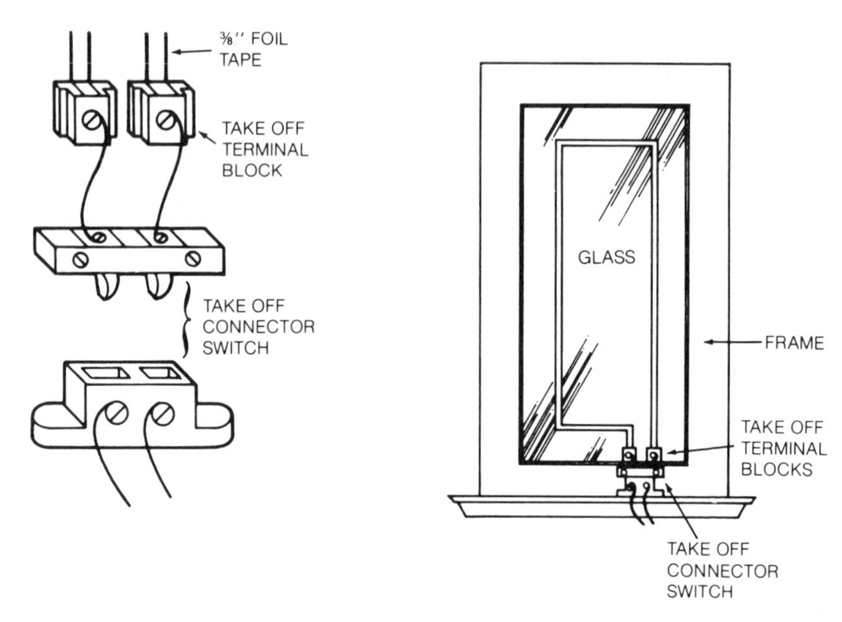

Figure 4-4. *Taped Window*

Most of today's tape has adhesive already applied so all one needs to do is remove the back and press it into place. Older types of tape used varnish to hold them in place. Once the tape is in place, it is connected to special connectors blocks and a take-off connector forming a normally-closed switch *(Figure 4-4)*. If the window is opened or broken the alarm sounds.

If magnetic window switches are installed for a perimeter system and open-window ventilation is desired, it can be provided by mounting an extra magnet below the switch as shown in *Figure 4-5*. If the window is raised so that the second magnet occupies the location shown for the first magnet, the system can be armed with the window open. This method may be easier than providing a special on-off switch as discussed earlier *(Figure 3-18)*. In either case, security is lowered somewhat. An intruder now has some limited access through the window which he might easily use to his advantage.

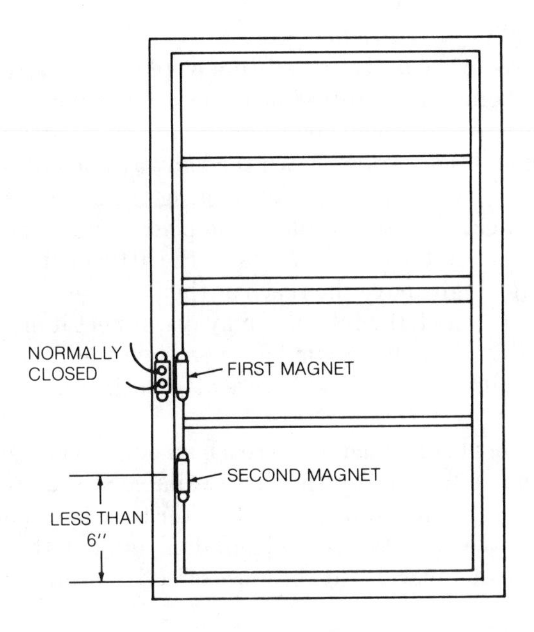

Figure 4-5. *Window Prepared for Ventilation*

As mentioned previously, the same features of a hardwired perimeter system can also be provided on a wireless perimeter system or one using the power lines within the home. The additional costs of the wireless system's hardware must be compared against the installation cost of a hardwired perimeter system to determine which is the most cost-effective solution.

Space Detection Systems

A Microwave System

A microwave space detection system would also offer good protection for this home. A suggested placement for two units is shown in *Figure 4-6*. Although all openings are not covered, an intruder would undoubtedly travel down the hallway and enter the living and dining area at some time while looting the home. The units should be connected to an outside alarm placed under the eaves of the home or, as shown in *Figure 4-7*, inside the attic near ventilation openings high above the reach of the intruder. If the siren or bell can be reached, the intruder may disconnect it or render it useless in some way (shaving cream has even been squirted into bells making them almost soundless) before attempting entry into the home.

Few intruders bring a ladder to reach an outside alarm, but some have been known to use ones that are available at the property. A careful radiation pattern test of the units after they have been mounted on firm stable mountings and behind paper or cloth screens will determine the exact placement for most effective protection.

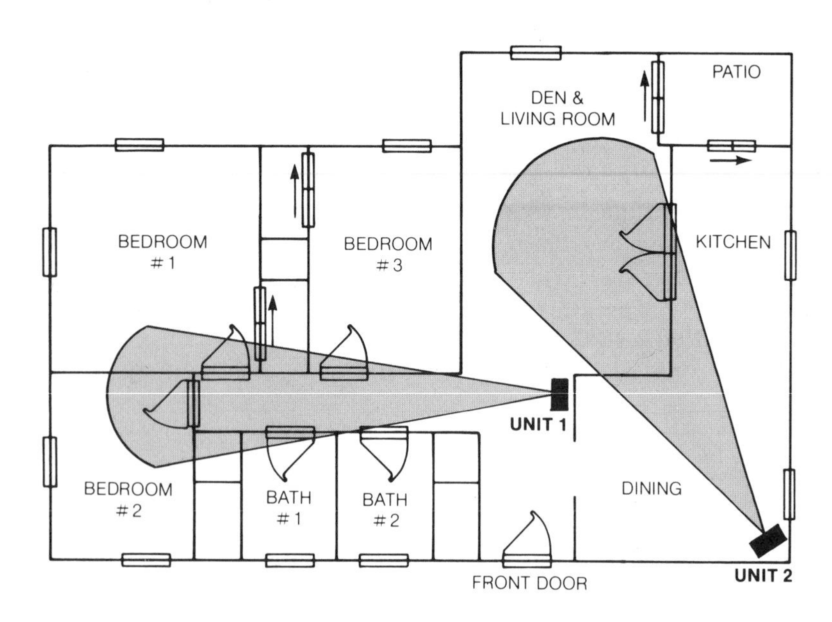

Figure 4-6. *Microwave Protection*

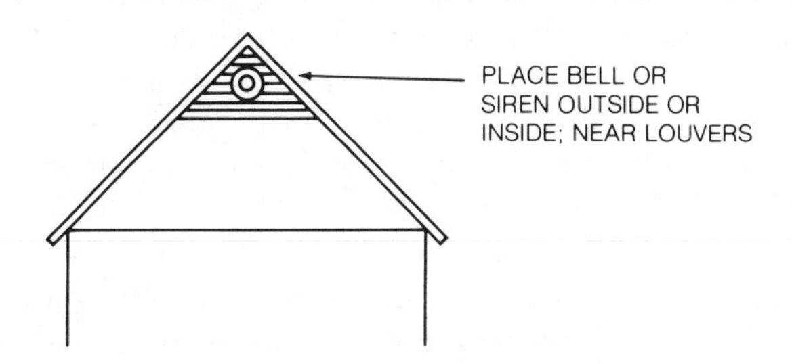

Figure 4-7. *Placement Of An "Outside" Alarm*

An Ultrasonic System

Two ultrasonic units could be used instead of the microwave units described but with a reduction in coverage due to the lack of penetration through walls into other rooms. However, there is another installation of ultrasonic units that can be more effective. For this type system, there are a number of *detect* and *communicate* channels and one *act* output as shown in the system diagram of *Figure 4-8.* Distributed transceiver units which receive their power from a central supply in the control box radiate individual detection patterns at their respective locations. They send their detection signals to the control box to sound the alarm when an intruder disturbs their field. Each one is controlled with an ON-OFF switch at the central control box.

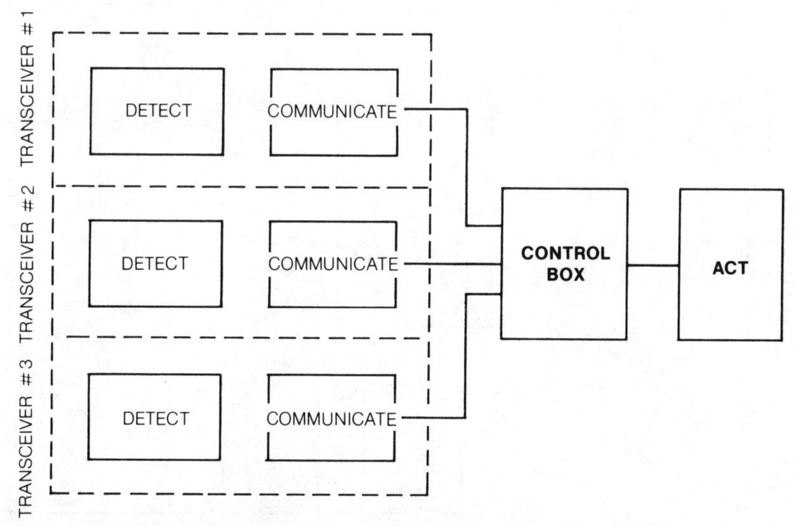

Figure 4-8. *Ultrasonic Detection System With Distributed Transceivers*

One of the transceiver units is shown in *Figure 4-9*. With such units it is possible to have a radiation pattern cover each room and give perimeter protection with a space detection system. For example, *Figure 4-10* shows the installation of six transceivers to provide the detection system for the home of *Figure 4-6*. Each room transceiver is providing protection for the external openings into that room. Thus, anyone entering a window or a door would cause the alarm to be sounded. With such coverage, it is unlikely that the intruder would go beyond an attempted entry into the home or business.

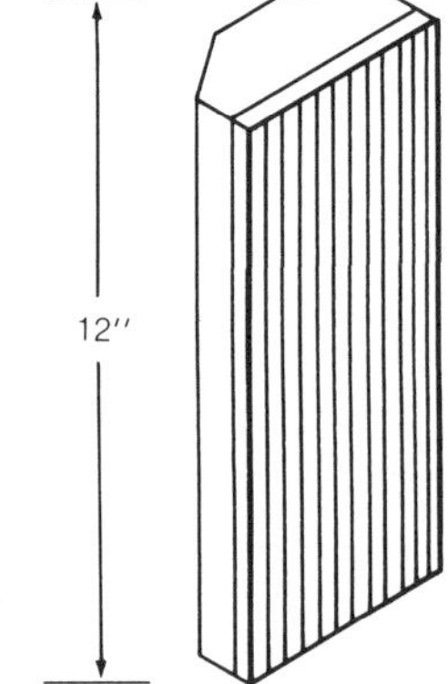

Figure 4-9. *An Ultrasonic Transceiver*

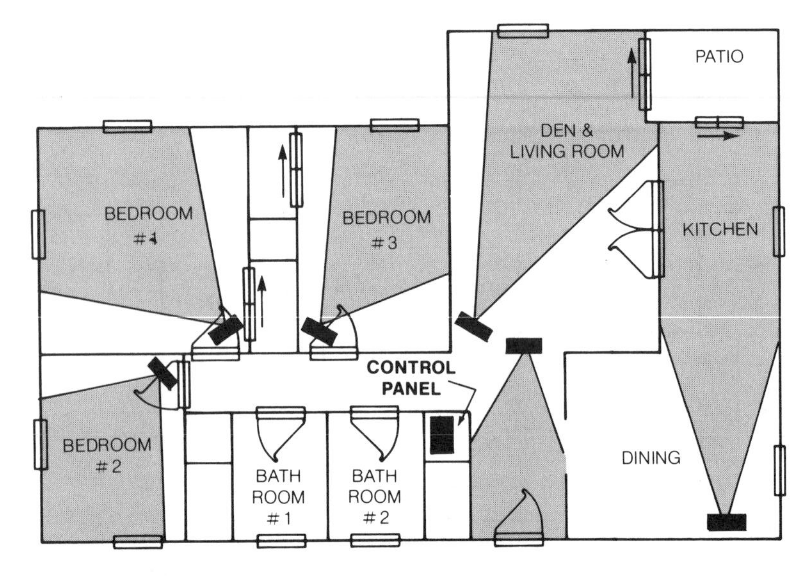

Figure 4-10. *Protection With Individual Transceivers*

Transceiver Installation

Transceivers can be installed in several ways. They can be mounted horizontally, vertically, butted in a corner, or at an angle from the ceiling *(Figure 4-11)*. Feet and frames are available to set the unit on a bookshelf or to install the unit flush with the wall. The transceivers in *Figure 4-10* are mounted to the ceiling of each room. The electrical power and control wires go through the ceiling to the attic. The corners of the ceiling near the center of the house are most likely to have ample attic space above them for easy installation access. This is where the control box is installed. Each transceiver connects to the control box with an unshielded cable containing four wires, much like a telephone interconnecting cable. Each transceiver has a LED (Light-Emitting Diode) indicator to show when it is in an alarm condition. The control unit for this system also has LED indicators available to show which transceiver is in an alarm condition. It is advisable to hide the control box from sight so that it cannot be located and the system deactivated by the intruder.

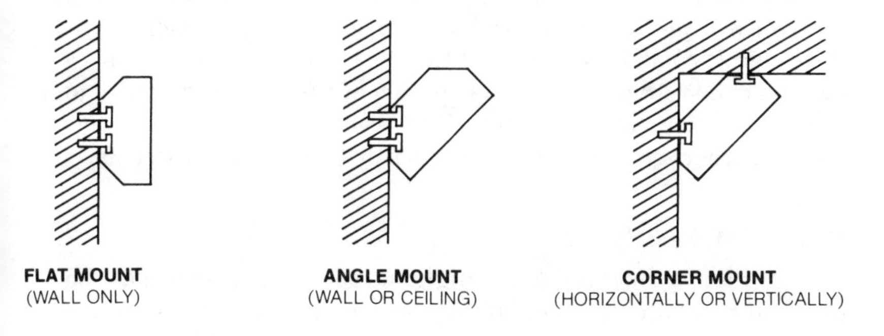

Figure 4-11. *Transceiver Mounting*

Transceivers are also available in dual units which extend coverage in large areas for little extra cost *(Figure 4-12)*. The total system cost of such a system is expensive compared to the simpler systems but it must be remembered that very complete dual coverage is obtained, both space and perimeter. Such units are particularly well suited for industrial installations. In fact, most system components have been designed for that application. To compare cost, a system with six transceivers and one control box presently costs approximately $1,000. Significant cost reductions are forecast for systems of this type in the future.

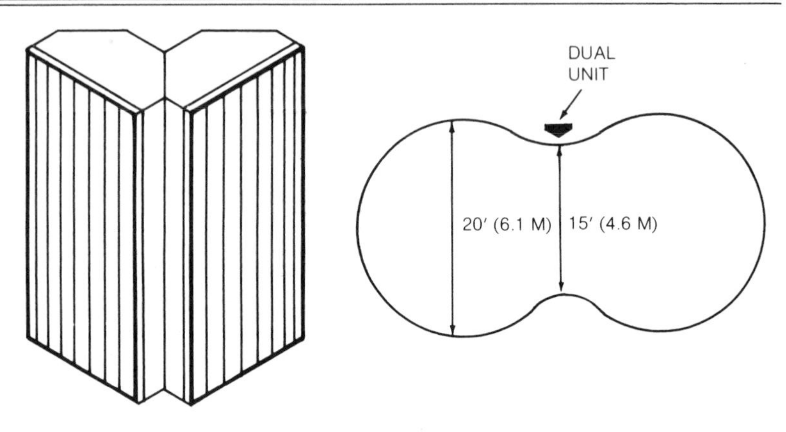

Figure 4-12. *Dual Ultrasonic Transceiver And Coverage*

Active Infrared Systems

An active infrared system might also be considered in this application. Recall that active means that it has a transmitter and receiver, not just a receiver, radiating a beam of light from one to the other. It could be used in addition to other coverage or as a stand-alone system. One suitable placement for the transmitter and receiver is shown in *Figure 4-13*. Units could be placed around the home in other strategic locations. Each system would cost about $250.

Passive Infrared System

Passive infrared units might also be installed much as the ultrasonic transceivers were in *Figure 4-10*. If only one unit was used instead of the multiple units in each room then a likely place might be the hallway leading to all the bedrooms. This placement is shown in *Figure 4-13*.

Passive infrared systems tend to lose sensitivity as temperatures rise above 90° F (32° C), therefore, during the summertime when a passive infrared detector is in use, some air conditioning should be available in the home or business. The unit should be placed so it will not receive direct sunlight from a window because direct sunlight reduces the unit's sensitivity.

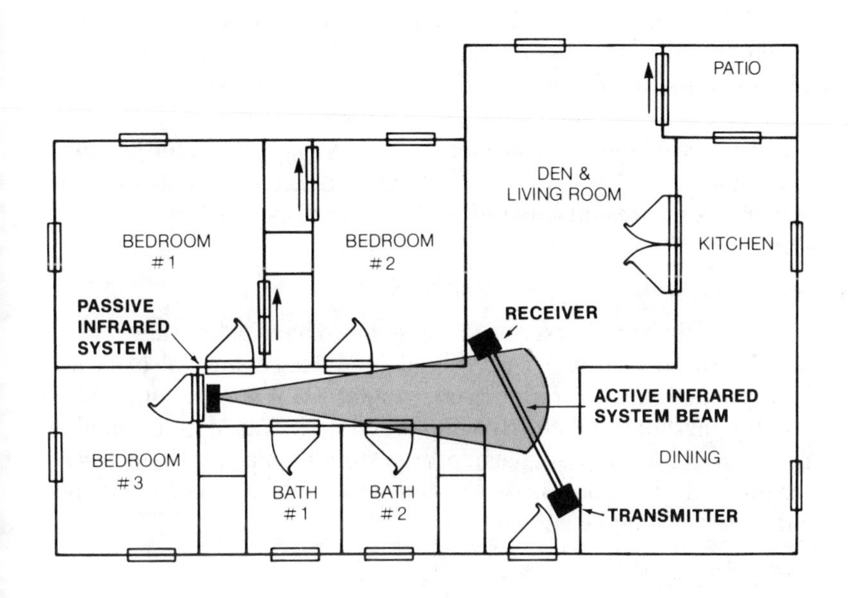

Figure 4-13. *Active Infrared Protection*

Which System Should One Choose?

Perimeter

All of the systems discussed have their merits and no clear-cut distinction exists. The hardwired solution, particularly if space protection is used as a backup, is a good system. When additional detectors are installed such as pressure senitive mats, temperature switches, and outside sirens it is probably one of the best. Wireless perimeter systems offer the same advantages, but fully installed the cost runs higher. Still, if the system is installed in one's spare time, it may be affordable. The perimeter systems have the advantage of also protecting the home from surprise entry while someone is at home.

Space

The microwave and ultrasonic space detection systems are extremely difficult for an intruder to foil. There are no switches or wires to locate and deactivate. Such units have the potential to false alarm under certain conditions but modern units properly installed have eliminated most of the common problems. Ultrasonic systems are even being designed which will not be as affected by turbulent air currents as present-day units. However, none work if there is a ceiling fan.

Active Infrared

The infrared systems discussed also have advantages. The active infrared system is very reliable and will seldom false alarm. The beam is difficult for an intruder to detect. With only one or two units the installation is fairly simple. Active infrared units could be used to create a complete perimeter system by placing them across each opening. Installation and expense would be extensive but the final system would be a good one.

Passive Infrared

The passive infrared system is fairly costly (one unit with a local alarm has an approximate cost of $300) but is extremely easy to install. This system has low power requirements so batteries often make it independent of home supply. These units are constantly being improved and may be a more viable choice for the home as the units are reduced in cost.

Technology is advancing each day. As particular systems improve over their present performance, one system may become superior to another in a clear-cut way. Therefore, before deciding on a system, consult a professional installer and listen carefully to what he has to say. It is likely that he has faced situations similar to the case at hand and is very familiar with local conditions. If the system that is recommended doesn't seem reasonable, get another opinion; as the old saying goes – two heads are better than one.

CASE 3

Most of the past examples have been residential. Case 3 is a small retail business which has a large glass frontage. It needs a protection system. The store is located in a small shopping center with lots of other shops the same size. As shown in *Figure 4-14*, the sides of the store are adjacent to other small shops which are assumed to be secure. The back door is a large metal one with a heavy steel bar across it, making entry with anything less than an acetylene torch very unlikely. The main concern is that someone will break the front glass and enter. Because of showcase windows and special decorative techniques it is assumed that the windows cannot be taped for a hardwired perimeter system.

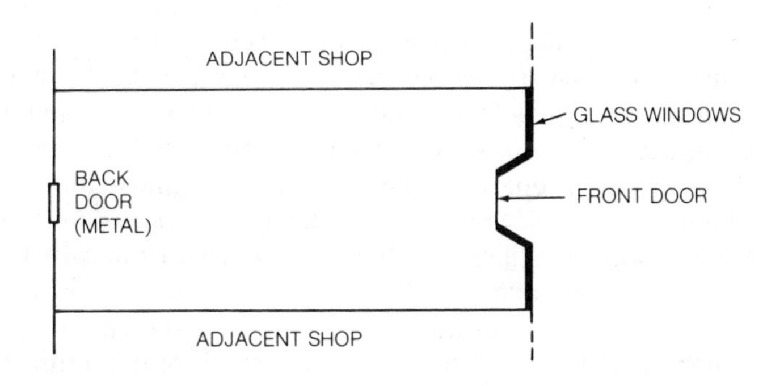

Figure 4-14. *Floorplan Of A Small Retail Shop*

Sound System Solution

A good solution to this security problem would be a sound discriminator which detects the noises generated in a break-in – particularly the sound of breaking glass. The system might be connected to a local alarm as well as to a unit that dials phones from remote locations. Such a dialer would call the police (if they allow direct hookup) and call the owner or manager at home. This system could be installed as shown in *Figure 4-15.*

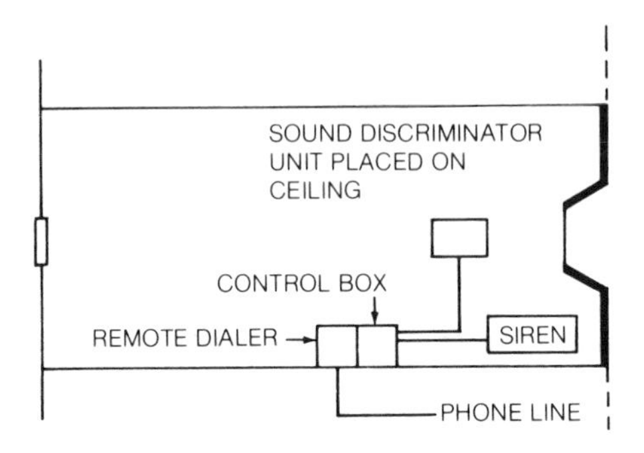

Figure 4-15. *Sound Discriminator System*

There are two basic types of remote dialers, tape and digital. Tape dialers use a tape recorder playback to call one or several numbers and give a voice message. Digital dialers electrically generate a series of tones which identifies the location of the alarm to a specially prepared remote station. For this example, a tape dialer is chosen because continuous monitoring is not required. A remote station which monitors a system continuously, though desirable, often adds a cost of a minimum of $20 per month.

A block diagram of a tape dialer is shown in *Figure 4-16.* When an alarm condition exists as a result of detection, the control box signals the tape player by supplying a voltage which starts the player. The tape player first generates the tones used to dial the number selected and then plays the voice recording. Tones can also be used to "hang up" the phone and another number may be called in the same manner. As many numbers may be called as are preprogrammed on the tape. Generally the tape is made in a continuous loop with a metal clip placed at the end to turn the recorder off after the tape has played. The connection into the telephone line is provided by the phone company at a nominal charge.

The sound discriminator system with the tape dialer will range from $400-$500. A special piece of equipment, either available where the tape dialer is purchased or at an electronic detection system sales office, must be used to record the closed loop tape with the phone numbers and messages desired.

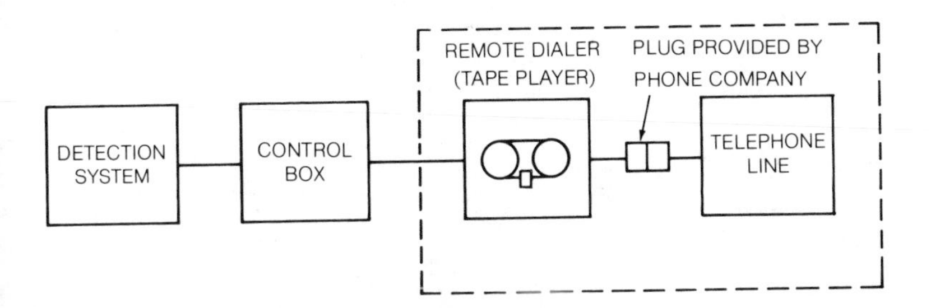

Figure 4-16. *Block Diagram Of A Tape Dialer*

Ultrasonic Space System

The next best alternative detection system would probably be an ultrasonic space detection system similar to what has been described previously. The alarm system can be coupled to the same tape dialer for the same type remote alert. System cost will be approximately the same.

CASE 4

Case 4 is another industrial application. This time the application concerns an appliance store. The store offers washing machines, dryers, and furniture for sale at retail. Business has been good lately so the owner has leased some extra storage space to accommodate several train loads of inventory that have been purchased at a special price.

Figure 4-17 shows the back part of a large storage area that has been rented to store the inventory. The items are to be stored on skids which are placed in large racks with aisles between them so loaders and unloaders can drive through with forklift trucks. There is a large metal door at the end of the storage area for shipping and receiving goods. The truck loading door locks, but there is some indication that if heavy equipment were used, the door could be pried open. Thousands of dollars have been invested by the owner in this inventory and he doesn't want to lose it. What sort of alarm system would be best suited for this application?

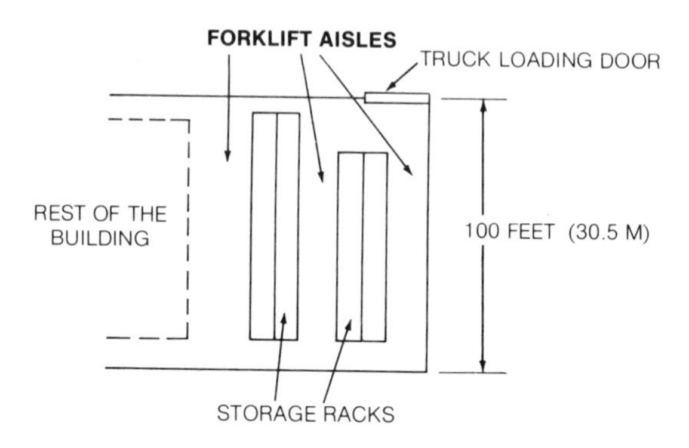

Figure 4-17. *Storage Area Floorplan*

The assumption is that the warehouse space will not be rented permanently. Therefore, a minimum amount of money is to be spent installing the system. However, a similar type space may be rented at another location later and it would be desirable to protect future locations with the same equipment. In addition, the storage space is not air conditioned, therefore the temperature and humidity change with the weather. Upon inspection, it was noted that there is considerable air turbulence generated when the wind blows against the loading door. The conclusion then is that a space detection system would be the best type system to use, and that the changes in temperature and humidity would cause the coverage of an ultrasonic space protector to vary and be unreliable. Besides that, the wind turbulence might cause false alarms particularly if the unit were improperly positioned. The large amount of area to be covered along with the conditions mentioned makes microwave space protection the best choice.

Three industrial type space detectors with a range of 100 feet or more would cover the area very nicely as shown in *Figure 4-18*. The walls of the warehouse are constructed with stone blocks and there are no windows so penetration is not a problem for units 1 and 2. Unit 3 faces a potential problem. The metal door moves back and forth (rattles) when the wind blows; this vibration could cause unit 3 to false alarm. But there seems to be an easy solution to the problem – the unit is aimed down so the protection pattern ends on the concrete floor in front of the door. The pattern position can be determined by conducting a test sequence like the ones discussed earlier.

Although the protection patterns are not shown to overlap in *Figure 4-18*, most units would have a wider field of coverage. Some of the microwave signals from one of the units might reach another unit and generate a false alarm. Most manufactures have taken care of this problem by building 4 or 5 types of units which operate at distinctly different frequencies. The units are often marked A, B, C, D, and E. The overlap problem is solved by making sure that each of the systems operate at a different frequency.

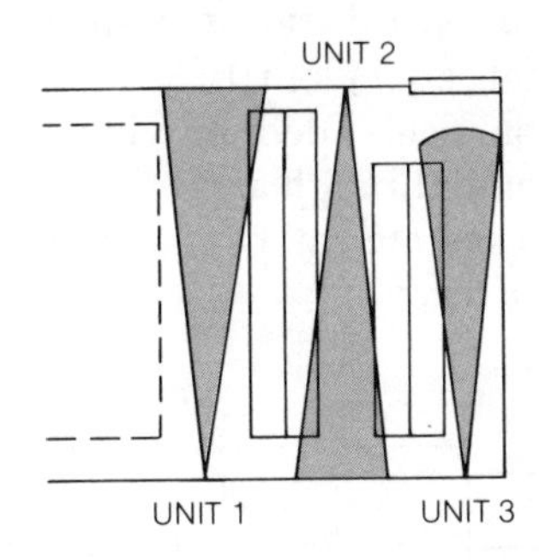

Figure 4-18. *Microwave Coverage Of The Area*

CASE 5

Case 5 is another business application; this time it is a manufacturing plant employing about 25 people. The company has accumulated expensive manufacturing equipment in the plant and management has been thinking of ways to protect it from robbery or vandalism. They have just about decided to have an alarm system installed, but find they also have other security related problems which might be solved at the same time.

The company does not employ a night shift in the plant, but occasionally employees work at night for some special reason. Management wants to maintain a "family type" of relationship between the company and its employees, but they feel they must monitor all after hours traffic in the plant. They want to know who enters and when. If there are any problems the next day they will know who to ask. Also, there are several areas within the plant that are not to be entered at night, e.g., certain supervisors are the only ones who are to have access to the model shop and the front offices.

Can an electronic detection system take care of their needs? After calling several alarm installation companies, the facilities manager is directed to a company which not only installs alarms but also maintains a central station.

The Central Station

Basically, a central station is a method of electronically watching a large number of businesses and residences. The individual alarm systems from the remote locations are connected by digital dialers and telephone lines to the central station. One system is shown in *Figure 4-19*. Several connections are shown between the alarm control box and the digital dialer. This indicates that several different types of alarm conditions may be communicated from a home or business to the central station with one digital dialer. These messages can (in digital code) identify if a violation has occurred in a particular area, if a particular door or window has been opened, or even if someone has come through a gate or onto a driveway. In other words, the alarm system can be coded to react in different ways to different violations.

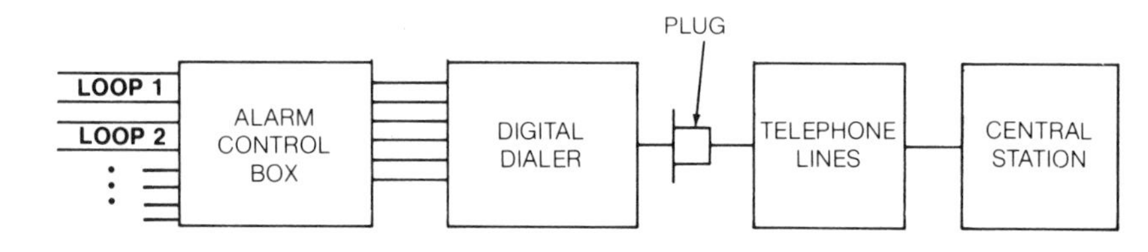

Figure 4-19. *Alarm System Connected To A Central Station*

Besides sending out the alarm code, a local siren at the plant can be connected to sound when certain types of violations have occurred. For instance, a local alarm is sounded if an outside window is broken. Once the alarm code is sent out through the digital dialer it is received by central station equipment which decodes the digital message and prints out the identifying number of the location, and the type of violation. The time of the occurrence is also printed.

Once the message is received the person who is monitoring the central station will check the type of violation and take the appropriate action. If the alarm condition is a violation which reflects a break-in, a security officer would be dispatched to the location and/or the police notified. Other alarm conditions may be handled in a different way. For example, if the front door is opened, the central station monitor may be instructed to wait five minutes for a phone call from the person entering. The person entering may be instructed to identify himself using a special number and then to give the reason for his entry.

The central station will also receive a coded message when the detection is turned off or on. If it were turned off at 8:00 in the morning no action would be taken; but if the time were 12:00 at night action would be taken.

The phone lines which connect the digital dialer to the central station are often called leased lines because they are designated for a special purpose. Sometimes alarm conditions are coded to indicate if a leased phone line is cut. Other features include digital dialers that have the capacity to seize a regular phone line even if it is busy.

This summary in no way includes all the services that are available from a central station, but it gives an idea of the types of services offered. The security company operating the central station may offer patrol services as well. These services do not come free of charge, but the monthly fee may be well below what it would cost the company to do it themselves.

Design Specifications for a System

To understand more clearly how one might decide on a specific system, let's generate for this case a set of design specifications. The floor plan is shown in *Figure 4-20.*

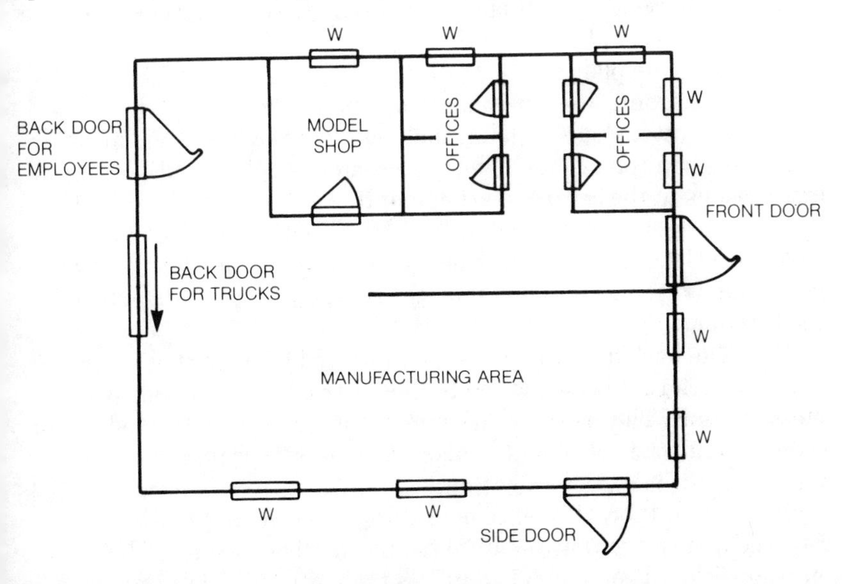

Figure 4-20. *Plant Floorplan*

Design Specifications

1. Provide perimeter protection for all openings.
2. Provide backup protection in key production areas in case an intruder were to elude the perimeter protection.
3. Provide a means to identify any employee and the time when that employee enters or leaves after normal working hours.
4. Provide security for the model shop and offices so that only selected supervisors can enter at night.
5. Provide monitoring by a security company which must dispatch a patrol car and make a personal check when a major violation occurs.

Designing A System

With the specifications in mind, the following security system is recommended. Outside windows of the building are to be taped and connected into a hardwired perimeter system forming a closed loop with the control box. When this loop is violated, a code 1 message is communicated to the digital dialer and the condition communicated to the remote station. Code 1 requires emergency action by the security company to investigate the violation. They are to call the police and dispatch their own patrol. Code 1 also sounds a local siren at the plant.

The front, side, and back doors are connected into a perimeter system loop with normally-closed switches. A violation would reflect a code 2 message to the remote station. Code 2 instructions to the central station monitor are to wait one minute for an identifying phone call before taking action on this type of violation. The reaction is the same as for code 1 if no identifying phone call is received, except only the security patrol is dispatched to the location.

The back door for trucks is connected by a magnetic normally-closed switch that is hardwired to the control box and a code 3 message sent. The central station monitor again is instructed, as for code 1, to call the police and to dispatch a patrol car immediately in response to this type of violation.

Central security office operators will have previously received a message that the total facility has been armed. The message "closed" or "open" identifies the status. The location or facility may be identified by a number, a code or a word, and even specific zones might be identified.

The model shop and office doors are hardwired in a separate perimeter system loop and any violation sends out a code 4 message. Code 4 also calls for emergency action. However, if no code 2 signals have been received, police would be called. With other employees already in the plant (code 2 signals received), only a security guard would be sent when code 4 is received because someone may have entered by error. If a code 2 has been received and a supervisor has identified himself as being on the list of permitted supervisors for the office or model shop, the code 4 is disregarded. Back-up support systems are provided by microwave space detectors with entry delay. These are placed inside the manufacturing areas where they can be switched off by an employee entering the area after hours and switched on again when the employee leaves.

The general layout of the detection system is shown in *Figure 4-21* which shows the overlap of the microwave systems onto the hardwired perimeter systems.

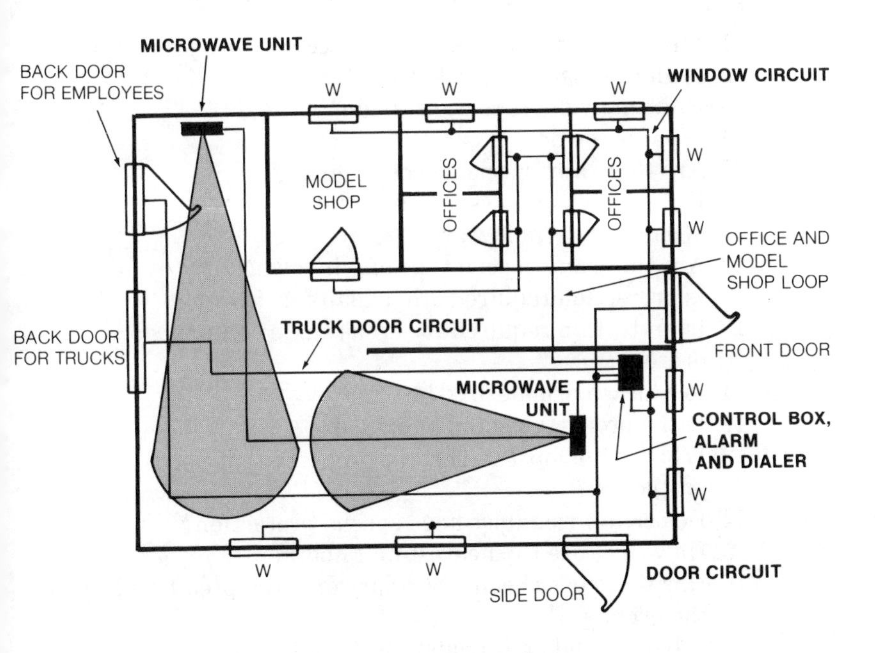

Figure 4-21. *Diagram Of Alarm System*

SPECIFICATIONS FOR OTHER SYSTEMS

Similar procedures can be used to set down specifications for other applications and the following checklist and instructions have been prepared to make this easier for the person that is interested.

Selecting a System

The detection systems shown in the preceding chapters have covered a wide range of applications. Anyone of these may be exactly what's needed. Frequently, however, they do not fit exactly. Each home or business is a special case and all conditions set up in the examples may not match. To aid in considering all the factors when choosing a system and to make sure that no steps are omitted, the following list of questions has been prepared. Some do not apply in all cases; ignore the ones that are not pertinent and add others that seem as important.

CHECK LIST

1. What is the total area of the space needing protection?
2. Is all the space on one level?
3. How many doors to the outside?
4. What is the door area?
5. Are there any special types of doors?
6. How many windows to the outside?
7. What is the window area?
8. Are there any special type of windows?
9. Is protection required other than for intruders?
10. Is protection required for areas while occupants are present?
11. Is a local alarm desired?
12. Is the property rented or owned?
13. What is the most important property that needs protection?
14. Do certain valuables need special protection?
15. How close are the nearest neighbors?
16. Has there ever been a burglary or attempted burglary on the premises?
17. Where would an intruder most likely enter?
18. Are some security devices already installed?
19. Is the area air conditioned in the summer?

20. Are smoke detectors present?
21. Are children often on the premises?
22. Is a handicapped person an occupant?
23. Are pets given free run of the location?
24. How old is the building?
25. How far to the nearest police station and fire station?
26. Are there budget limitations?
27. How many people are in a two-mile radius from the building?

The answers to the questions should aid in developing a set of specifications for the system required. Next a floor plan of the building to be protected should be drawn showing the physical locations of exterior openings and interior and exterior walls separating the areas to be protected. With the two – the specifications and the answers to the questions – it should be possible to get a preliminary idea of the kinds of systems and the number of systems required to meet the specifications.

Armed with the information above, a reliable distributor of security products should be consulted. He will be able to aid significantly by providing valuable system information and suggestions about specific systems. From these inputs, several alternative methods of providing the security required versus cost should result. One can then choose the system that is most cost-effective for the application.

WHAT'S NEXT

Now that particular cases have been discussed for separate systems and these have been combined in some cases into more complicated systems that supplement, complement or overlap each other, the discussion now shifts to more specifics on the control of these combined systems, both through a system element called the control box or through computer systems.

Quiz for Chapter 4

1. Most homes and businesses are robbed when:
   a. Only a few employees are present.
   b. When no one is present.
   c. During lunch.
   d. When only supervisors are present.
2. Which system is likely to penetrate interior walls of a home or business?
   a. Ultrasonic.
   b. Infrared.
   c. Microwave.
   d. All of the above.
3. When an alarm is placed in the "test" condition:
   a. The alarm sounds in a few minutes.
   b. The alarm does not sound at all.
   c. A test light comes on as the alarm sounds.
   d. None of the above.
4. The "tape" used to protect a window or glass area is:
   a. A nonconductor.
   b. Reflective tape.
   c. A conductor.
   d. Regular electrical tape.
5. Using several ultrasonic transceivers in a home offers the following disadvantage:
   a. Both perimeter and space protection are offered.
   b. Only one control box is used.
   c. Cost is high.
   d. The system is unreliable.
6. Passive infrared units should be installed so that:
   a. Temperatures do not exceed 90°F (32°C).
   b. Sunshine does not reach the unit.
   c. Dogs and cats do not enter the protected area.
   d. All of the above.
7. A sound detection system may be a good choice when:
   a. An intruder is likely to make noise when entering.
   b. Protecting a small area.
   c. When fans are present.
   d. When protection is needed only on holidays.
8. A ________ space detector should be used when temperature and humidity changes are severe.
   a. Ultrasonic.
   b. Microwave.
   c. SONAR.
   d. Seismic.
   e. Hardwired.
9. To avoid false alarms when space protection overlaps, manufacturers often:
   a. Use different frequencies.
   b. Suggest another brand.
   c. Suggest a metal shield.
   d. All of the above.
10. Special security measures often include:
    a. A remote dialer.
    b. Use of a central station.
    c. Both perimeter and space protection.
    d. All of the above.

(Answers in back of the book)

System Details and Computer Systems

ABOUT THIS CHAPTER

The discussion in previous chapters has focused on installation and selection of personal and business security systems. Now the discussion shifts back to the general systems so that more technical details can be covered that will lead to an understanding of how and why computers are coupled to security systems. In several systems that were discussed previously, the heart of the system was the controller or control box. Our discussion begins there.

THE CONTROL BOX

The control box of a modern electronic detection system is designed to:

1. Recognize the *detection* signal of the system.
2. *Communicate* that detection into an *action* that sounds an alarm or in some way announces or demands a response. As added features the design specifications might require the unit to:
   a. Keep the current requirements to a minimum for the detection devices, thereby, saving power. With the reduced current the system can use lightweight wire and switches, thereby, reducing the cost.
   b. Use normally-closed switches for the major detection loops but also allow use of normally-open switches.
   c. Provide entry and exit delays.
   d. Continue sounding the alarm even after the violation has ceased.
   e. Provide an automatic cutoff after the alarm has sounded for five minutes.

Frequently, the design also requires a battery backup so the system can operate during a power failure. For maintenance or test, a light is often required that shows a detection before the alarm siren or bell sounds. As a result, the system can be deactivated before the alarm sounds in case of a known false alarm.

Most of these features are made possible by the design of the circuits in the control box. Many different types of circuits are possible to accomplish the function desired. The circuits chosen for discussion may not be the exact circuits used in every controller, but they are representative of the basic concepts used to provide the features.

Reducing the System Material Cost

The simple alarm system of *Figure 2-2*, which consists of a power supply, a normally-open detection switch and a siren, is shown again in *Figure 5-1*. All circuit elements, the switches and the wire must be heavy enough to carry all the alarm siren current. With a simple one-switch circuit this is easy to do, but when several switches are used in different locations, the cost and bulk of heavy-duty switches and wire becomes a problem. It is much easier to conceal light-weight wires and switches.

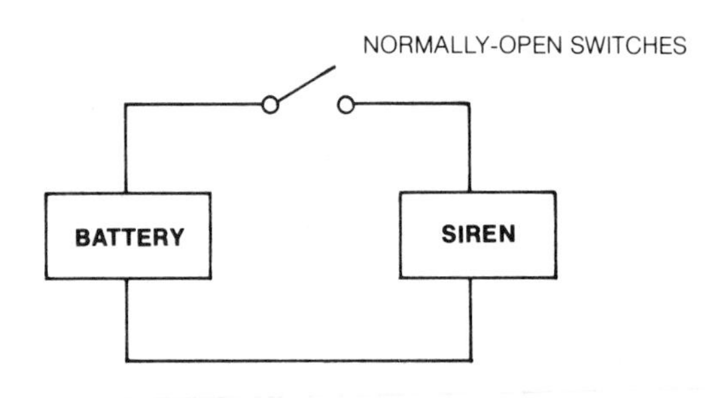

Figure 5-1. *A Simple Alarm*

Figure 5-2 shows again the simple alarm circuit of *Figure 2-4*. It protects three openings with *normally-open switches*. The current which the wires and switches must carry to sound the alarm can be reduced by using a relay in the control box.

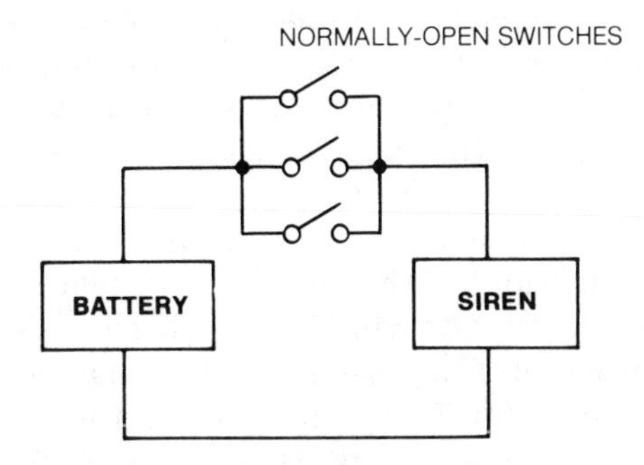

Figure 5-2. *Protecting Three Openings*

What Is A Relay?

When several loops of wire are wrapped around an iron bar, and a current is sent through the wire, the iron bar becomes an electromagnet. The resulting magnet can be used to pick up nails *(Figure 5-3a)* because the magnetic field attracts the iron nails. In like fashion, a set of contacts can be attached to a metal armature, but insulated from it. When the metal armature is pulled to the magnet the contacts close and complete an electrical circuit so that current can flow. This is the principle concept of a relay *(Figure 5-3b)*. Such a relay is used to switch the large battery current required by the starter of your automobile. This component of an automobile starting system is called for starter solenoid.

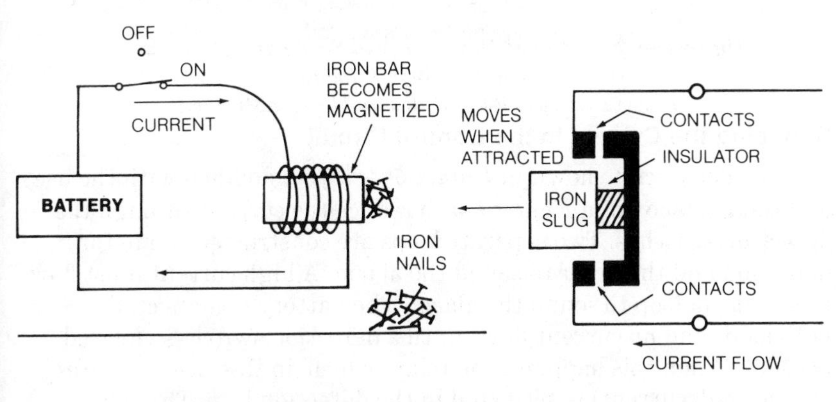

Figure 5-3. *An Electromagnet*

Figure 5-4 shows a functional diagram of a two-circuit (two-switch) normally-open relay, a schematic representation and a picture. The iron core and the contacts are mechanically placed together on the relay so the electromagnet can open or close the switches. Some relays have the contacts normally closed so that they open when the relay is energized, some have them normally-open, others have combinations of the two. For ease in drawing circuit diagrams, the relay contacts will often be shown in separate locations. This is because each set of contacts may be used in a different, separate circuit. Relays are manufactured that operate on a variety of current requirements. Their design depends on the number of contacts to be closed, the pulling power required, the physical size and the type of service required.

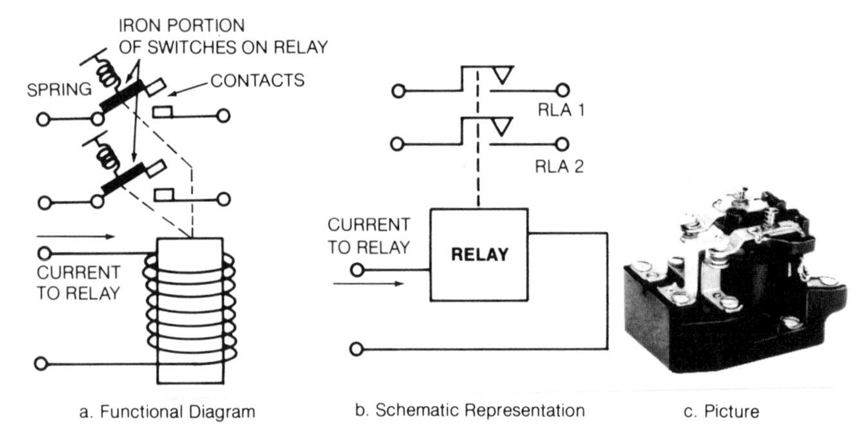

Figure 5-4. *Functional Diagram And Schematic Presentation Of A Two-Switch Relay*
(Photo Courtesy of Potter & Brumfield Div. AMF Inc.)

Reducing the Current in the Control Circuit

The circuit shown in *Figure 5-2* can be modified with the use of a relay as shown in *Figure 5-5* to reduce the current through the detection switches. Two separate loops are constructed – one for detection and the other to sound the alarm. A high current must flow in the alarm loop to sound the alarm. The battery is connected to both loops and no current flows until a detection switch is violated (closed). When this happens, the relay (which, in this case, requires only a small current) is activated in the detection loop. The relay closes the contacts RLA 1 and the high-current loop is completed sounding the alarm. Anytime any of the three detection switches are closed the relay is energized and the alarm is sounded.

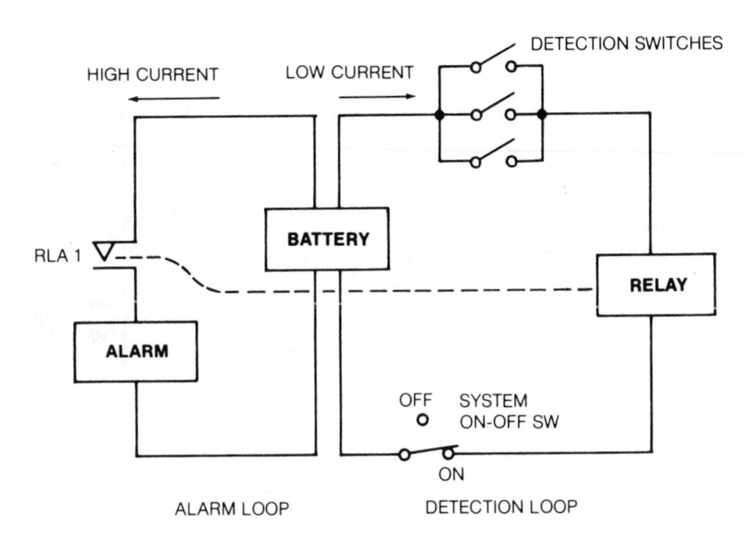

Figure 5-5. *Reducing Current Through The Detection Switches*

Some Other Relay Functions

Relays have other uses in a detection system. They allow electrical systems to be isolated from each other. For example, the low-voltage battery power used for the control box can, through the relay, turn on (or off) devices which require 110V alternating current power used in the home. For instance, a relay which energizes three sets of contacts could be used to sound an alarm operating from a battery system, turn on several 110Vac lights and start a tape recorder. The system is shown in *Figure 5-6*. When one of the detection switches is closed, current flows through the relay closing RLA 1, RLA 2, and RLA 3. RLA 1 sounds the alarms, RLA 2 turns on the lights, and RLA 3 starts the tape recorder.

Relays can be designed with different power requirements for a vast range of applications. They could be used to turn on a kitchen range, an airconditioning or heating unit, a radio or TV, etc. They can have a very simple or very complicated arrangement of contacts. In this example, relays have been used to save material cost and reduce the power used in the total system by reducing the current required in the control circuits.

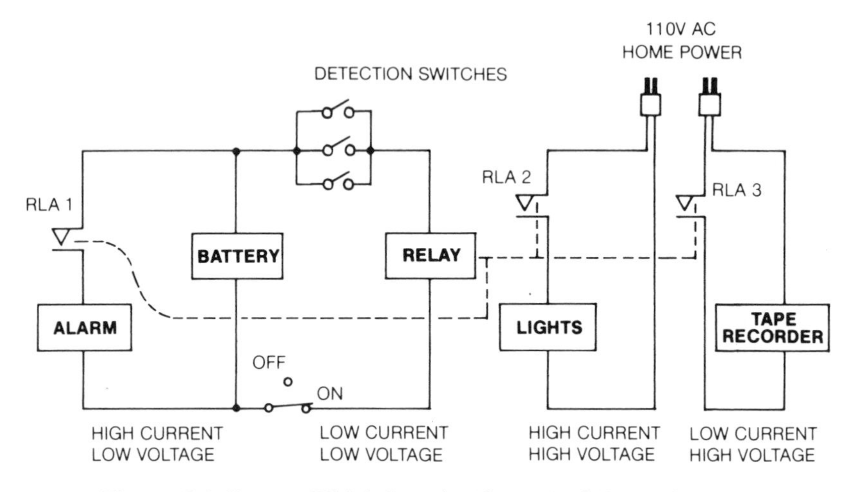

Figure 5-6. *System Which Sounds Alarm And Turns On Lights And A Tape Recorder*

Latching The Alarm

When an intruder enters protected premises, he may violate a magnetic switch by opening a window or door, but if the door or window is closed quickly, the alarm will stop sounding. The length of time the violation occurs may not be adequate to alert you, your neighbor or the proper security personnel. Extra circuitry in the control box can be designed which causes the system to "latch", holding the alarm condition. One simple way is shown in *Figure 5-7*. All that is necessary to add to the circuit shown in *Figure 5-5* is a relay with two sets of contacts. One set of contacts is added in parallel with the detection switches and the other set is used as before to complete the alarm circuits.

When one of the detecting switches is closed, the current flows from the battery through the detecting switch and the relay. This closes RLA 1 activating the alarm and closes RLA 2 which keeps current flowing through the relay even though the detecting switch is returned to normal. The alarm will sound until the system is turned off with the on-off switch.

In many cases solid-state logic devices have replaced the relay, performing the function in much less space, with much less power, faster and with more reliability. However, for circuit isolation and for high-power operation, the relay continues to be used and performs very well.

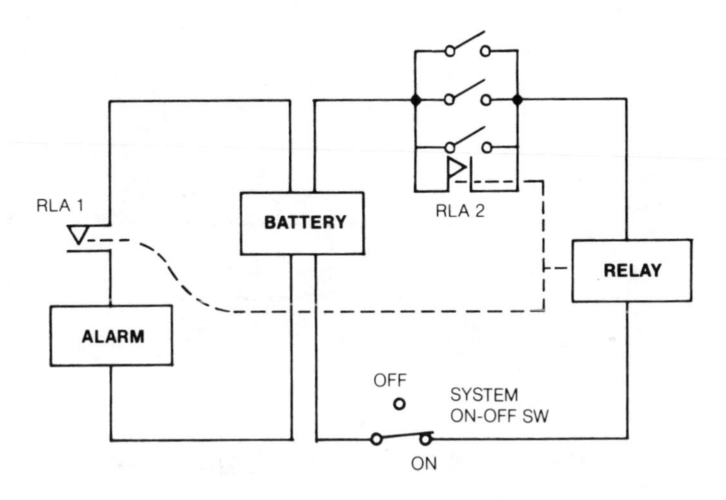

Figure 5-7. *Keeping The Alarm Sounding*

THE CONTROL BOX FOR A NORMALLY-CLOSED SWITCH SYSTEM

The normally-open condition of the detector switches that have been shown in the preceding examples have a glaring disadvantage which has been mentioned before. If a break occurs in the detector switch wiring, no indication of the fault condition is given. In such a system, an intruder may cut the wires without sounding the alarm. In the preceding chapters, systems were shown that overcame this drawback by using normally-closed switches in either a branch or throughout the total system. The control box becomes more complicated when a system uses normally-closed switches and will most likely include solid-state devices such as transistors, diodes and integrated circuits.

In order to understand this operation, a simplified explanation of a transistor used as a switch in a circuit follows. A transistor *(Figure 5-8)* has three connections which are called the emitter, base and collector. The one shown schematically in *Figure 5-8a* is an NPN transistor. A cross section is shown in *Figure 5-8b.* It is so named because it is made of a sandwich of solid-state materials that have been doped with impurities to make them N-type or P-type semiconductors to obtain transistor action.

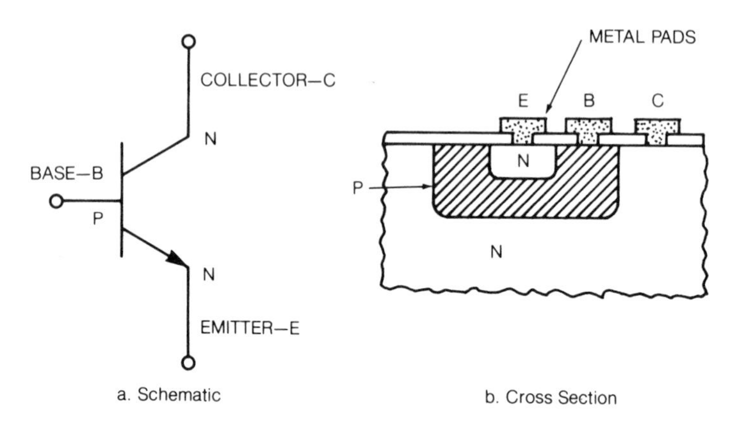

Figure 5-8. *Solid-State Device—A Transistor*

When the base voltage is 0.7V more positive than the emitter voltage the transistor turns on, i.e., current flows from collector to emitter as shown in *Figure 5-9.* The current from collector to emitter is controlled by a current from base to emitter. The ratio of collector-emitter current to base-emitter current is large, from 50 to 200, therefore very little current flows from the base to the emitter (for our purposes it will be neglected entirely).

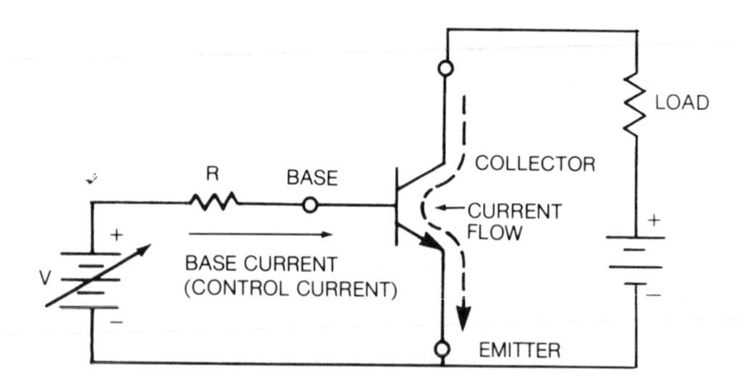

Figure 5-9. *NPN Transistor Conducts Current From Collector To Emitter*

Figure 5-10 is a portion of a detection system using normally-closed switches. When the normally-closed switches are closed some current flows in the loop as shown. The voltage at point A is zero because all of the battery voltage is dropped across resistor R1. Another way of looking at this is that the switches have a very low resistance when they are closed and, therefore, connect point A to the zero-voltage side of the battery.

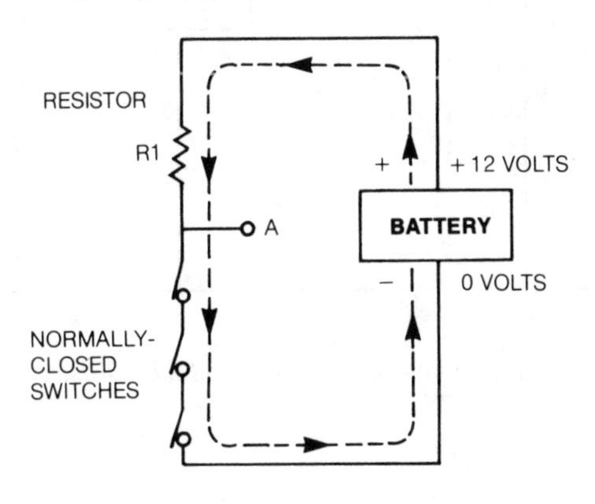

Figure 5-10. *A Portion Of A Normally-Closed System*

Now, if one of the normally-closed switches is opened, the voltage at point A will be 12 volts since no current can flow in the circuit and through resistor R1. Point A is effectively connected to the positive side of the battery. In summary, the voltage at point A is controlled by the normally-closed switches; when the switches are closed the voltage is 0, when one or more is open the voltage is 12 volts.

The circuit of *Figure 5-10* will now be used along with a latched circuit of the type used in *Figure 5-7*, except now the detection circuit will be using normally-closed switches. The total circuit is shown in *Figure 5-11*.

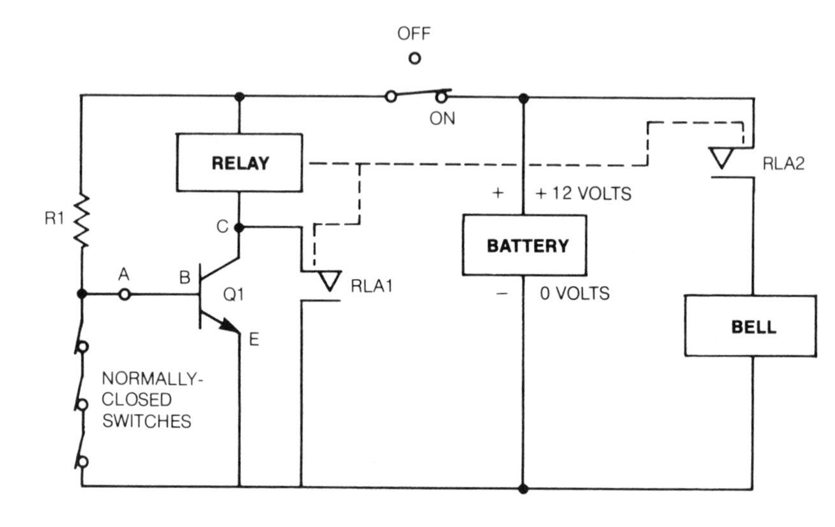

Figure 5-11. *A Normally-Closed System Which Latches*

System Operation

When one of the normally-closed switches is opened, the voltage at point A increases, causing transistor base-emitter current to flow which turns on the transistor. This allows current to flow from the battery through the relay (from collector to emitter of the transistor) and back to the battery. This current flow pulls-in the relay to close RLA 2 which sounds the alarm and RLA 1 which latches the relay. Regardless of the condition of the normally-closed switches the alarm current now continues. The addition of the transistor has provided the necessary cirucuitry to use normally-closed switches in the same fashion as the normally-open switches were used previously.

CONTROL BOX SYSTEMS WITH ENTRY AND EXIT DELAYS

For the systems discussed thus far, in order to turn on a system, the occupant would have to leave the premises and then turn on the system. The on-off switch would be outside the area of coverage and would have to be turned on by authorized personnel using a key. An extra key would have to be carried by the people who wanted to energize the system. If keys were lost or if personnel changed rapidly, then the security of these systems would be lowered accordingly.

THE DELAY CIRCUIT

Exit and entry delay circuits solve this problem. Let's look at the basic principle behind such delays. They are used in detection systems to delay the time when a system is activated or de-activated. This allows an occupant to leave or enter and not trigger a false alarm.

In *Figure 5-12*, when the on-off switch is turned on, current flows through R1, branches and flows through R2 and back to the battery. At the branch it also flows through R3 and into the capacitor C. The voltage on the capacitor starts at zero volts and becomes more positive as the current flows into it from R3. The transistor will not conduct or turn on until the voltage at point A is more positive (by 0.7 volt) with respect to the emitter. Diode D1 in series with the emitter adds another 0.7 volt between the emitter and the minus terminal of the battery so that point A must reach +1.4V before the transistor conducts. The current through R3 charges the capacitor slowly so that a time delay occurs between the time that the system is turned on and the time that the voltage at point A is positive enough to turn on the transistor, energize the relay and supply power (+12V) to the rest of the system through relay contacts RLB1.

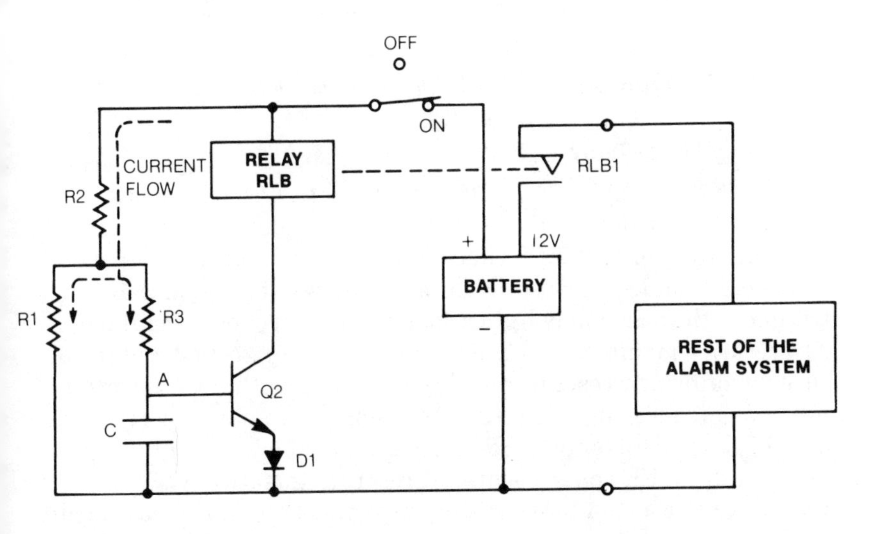

Figure 5-12. *Time Delay Circuit*

Combined Delays in a System

This same type of delay circuit can be used in the alarm circuit to delay the sounding of the alarm. Two of these delay circuits are shown in a normally-closed system in *Figure 5-13*. The exit time-delay circuit does not connect the battery to the detector circuit of the system until enough time has passed for one to leave the premises. The entry time-delay circuit in the alarm loop does not start until the system is in an alarm condition. It then delays the sounding of the alarm for a time interval sufficient for one to enter and turn off the alarm.

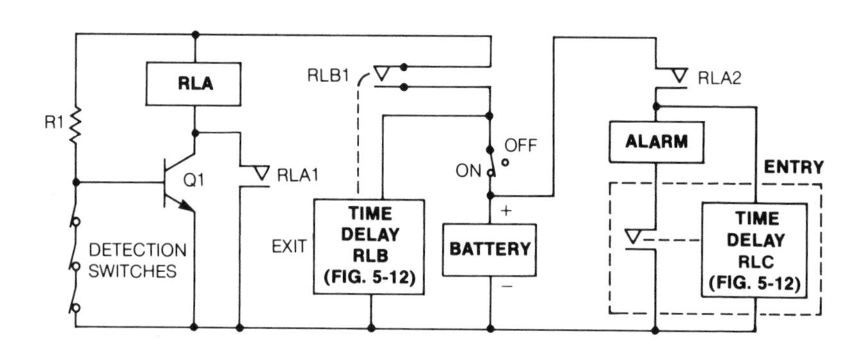

Figure 5-13. *System With Entry And Exit Delay*

Resetting Time-Delay

Some systems reset automatically after the alarm has sounded for five or ten minutes. After an intruder trips the alarm, he certainly will not remain around for five or ten minutes after the alarm has sounded because, as far as he knows, the signal that triggered the alarm may have also notified the police, a neighbor or a security monitoring station. Even so, the biggest advantage of such an automatic reset is probably in dealing with false alarms. If lightning strikes and triggers the system it automatically resets. If an occupant mistakenly triggers the system at a location far removed from the control center, it resets automatically. This feature can be added to the system using another time delay-circuit as shown in *Figure 5-14*.

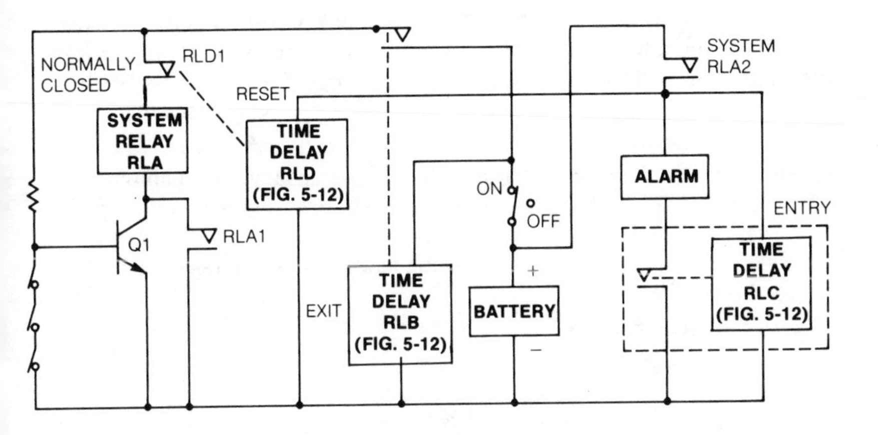

Figure 5-14. *System With Reset Feature Added*

When the alarm circuit is energized both the entry and reset time delays begin. The entry time delay allows the alarm to sound after 30 seconds, and the reset time delay would be set to act five or ten minutes later. The entry time delay allows the alarm to sound; the reset time delay turns it off. The reset time delay has a set of normally-closed contacts in the series with the system relay. When the RLD relay energizes, it opens contacts RLD1 and deactivates RLA which turns off the alarm and releases the latch. The system would not cycle back to an alarm condition unless the detection switches were still violated. If the system is still in alarm condition, the alarm will sound again and the time delay will cycle again.

The actual circuitry found in a time delay circuit may vary considerably. It may not contain a relay at all. It may use solid-state devices, even integrated circuits to provide the time delays and ON-OFF devices to replace the switch closures, but the basic operation is the same.

With the addition of each succeeding circuit the control box gets more complicated; however, by separating out each circuit and remembering the basic concept, the system is easily understood. Remember that most systems have provisions for the use of both normally-open and normally-closed switches and some have provisions for several separate branches of each.

CONVERTING 110VAC POWER TO 12 VDC

Detection systems typically utilize around 12 volts for safety reasons. Such a low voltage offers little danger of electrical shock or fire. This low voltage is obtained using a transformer and a bridge rectifier as shown in *Figure 5-15*. Some of the systems use filter capacitors to smooth out the 12 volts as shown in *Figure 5-15;* however, these filters are not absolutely necessary when charging a battery because the battery acts as a very large capacitor.

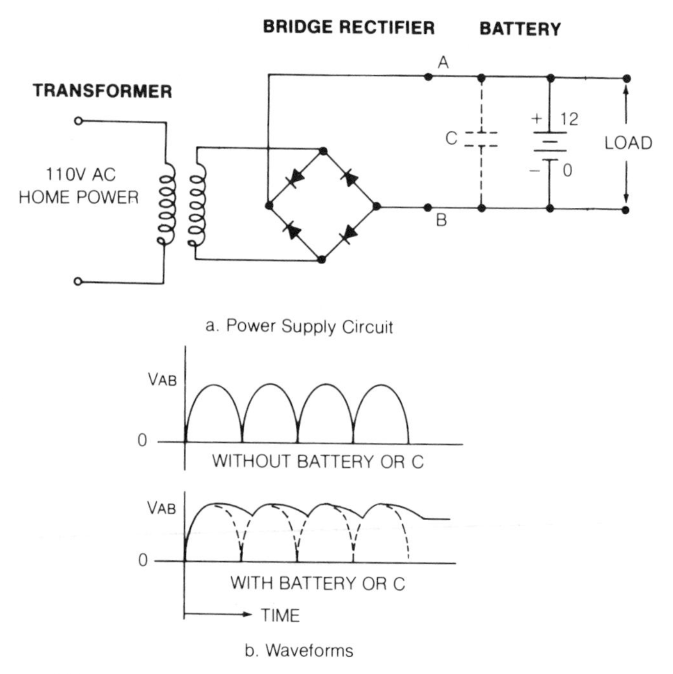

Figure 5-15. *Converting Home Power To 12 Vdc*

12 Volt Power Supply

The transformer reduces the home 110 Vac to a much lower alternating current voltage of around 17 volts. The bridge rectifier consists of four diodes which convert the reversing alternating current to a fluctuating direct current. The diodes conduct current only in the direction of the arrow head used in the diode symbol. The purpose of the bridge rectifier is to direct the current in the same direction through the battery to charge the battery no matter if the voltage applied is a positive or a negative half-cycle pulse (shown in *Figure 5-16.*).

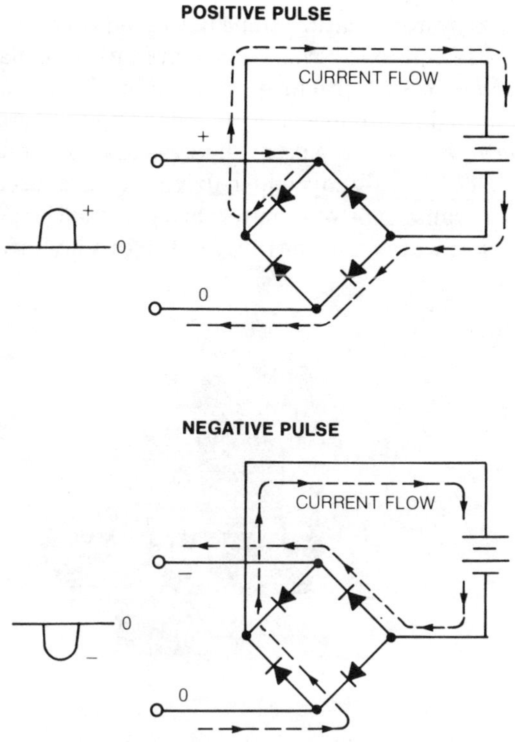

Figure 5-16. *Operation Of A Bridge Rectifier*

Notice that the current always flows in the direction of the diodes and from a higher to a lower voltage. In the case of the negative pulse, the current flows from the high voltage of zero volts to a negative (low) voltage. In the circuit shown, the battery is always being charged. Actual circuits often contain regulators which control the amount of current reaching the batteries and the voltage to which the battery is charged. Such precautions extend the life of the battery.

PROVIDING THE SIGNAL FOR A SIREN

The final system feature to be discussed concerns the alarm. Many systems use bells as alarms but a more effective alarm is a siren. Sirens *(Figure 5-17)* produce a much more identifiable tone and can be designed to produce a much louder noise. Actually, a siren is merely a high-efficiency speaker. The type of subsystem that generates the electrical signals which drive the siren uses computer-like circuits. The same type of circuits that provide the pulses used as "ticks" in a digital watch and to transfer and display information in a digital computer.

Figure 5-17. *Siren*

A siren needs a sound frequency (around 800Hz to 1200Hz – a *Hertz* is a cycle per second) signal to drive it. In the past the number of components required to do this made a siren circuit very expensive. Today, integrated circuits are available which provide a low cost method of providing computer functions to create the frequency and generate the signal required.

Computer Logic

The computer functions that are required are called decision or logic circuits. As shown in *Figure 1-9*, thousands of transistors, diodes, and resistors which form hundreds of logic circuits can be made together at one time in an integrated circuit. One that can fit on the tip of the finger. Such circuits are packaged as shown in *Figure 5-18* and provide the computer functions needed in detection systems at low cost.

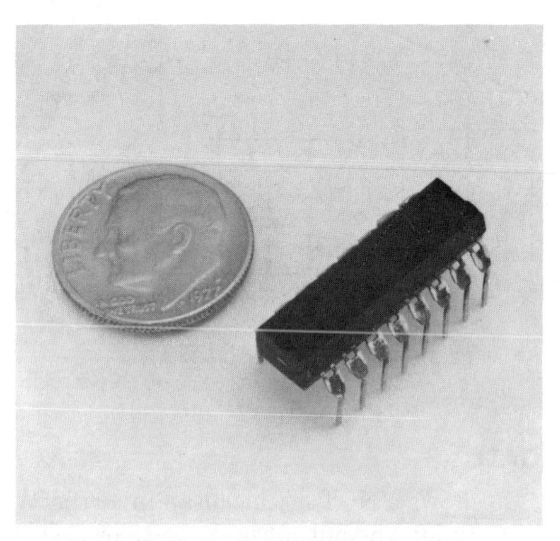

Figure 5-18. *Integrated Circuit Package*

Digital Signals

In a computer logic circuit there are only two levels used to identify information. A common way is to use a high-level positive voltage for one level (typically +5 volts) and a low-level voltage for the other level (usually 0 volts). To make it easier to use logic circuits and describe their operation in systems, abbreviations are used for the levels. H or 1 for the high level and L or 0 for the low level. These abbreviations can then be used to describe the behavior of binary logic circuits (they are called binary because two states or levels are used).

OR GATE

For example, *Figure 5-19a* is a table called a truth table. All possible states of the output relative to the signal states of the inputs are shown. This is an OR circuit (gate) and the symbol is shown in *Figure 5-19b.* The symbol represents the several solid-state components used to perform the function. The OR gate operates as follows: a high state appears at the output if either or both inputs have a high state. In terms of voltage it says: if there is +5 volts on one or both inputs the output will be +5 volts.

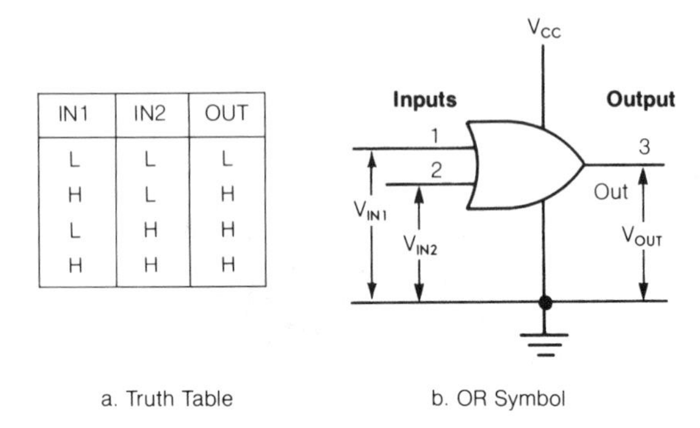

IN1	IN2	OUT
L	L	L
H	L	H
L	H	H
H	H	H

Figure 5-19. *An OR Gate*

NOT and NOR GATE

Figure 5-20 shows a NOT circuit or an inverter. Whatever state appears at the input, the output is opposite or inverted from it.

These two circuits can be combined into one called a NOR logic circuit *(Figure 5-21)*. Its output has states that are inverted or opposite to the OR circuit of *Figure 5-19* because there is an inverter at the output (this is indicated by the small circle).

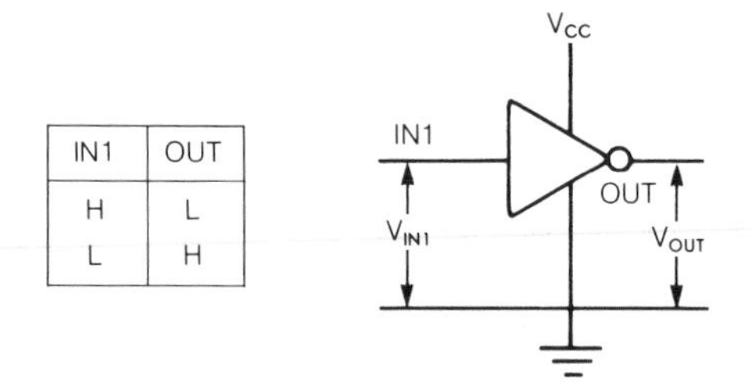

IN1	OUT
H	L
L	H

Figure 5-20. *A NOT Or Inverter Circuit*

IN1	IN2	OUT
L	L	H
H	L	L
L	H	L
H	H	L

Figure 5-21. *A NOR Logic Gate*

AND and NAND GATE

Another very popular logic circuit is the AND logic circuit. The truth table and its symbol are shown in *Figure 5-22a* and *b*. A high state only appears at the output if both inputs are in a high state. In like fashion to the NOR circuit, a NAND gate is formed by adding an inverter. This is shown in *Figure 5-22c* and *d*.

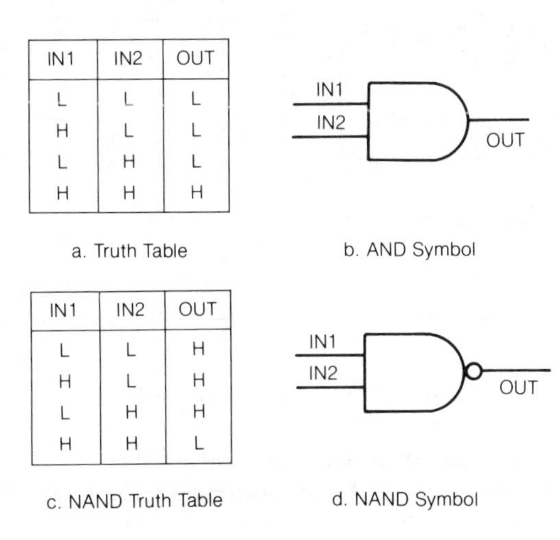

IN1	IN2	OUT
L	L	L
H	L	L
L	H	L
H	H	H

a. Truth Table

b. AND Symbol

IN1	IN2	OUT
L	L	H
H	L	H
L	H	H
H	H	L

c. NAND Truth Table

d. NAND Symbol

Figure 5-22. *AND and NAND Logic Gate*

Signal Generator for the Siren

Generating the signal for the siren may be accomplished by using NOR logic circuits and some additional components. The total circuit is shown in *Figure 5-23*. However, to understand the total circuit operation each of the two parts will be discussed separately.

Previously, an NPN transistor was used to show how a small amount of base current can control a large amount of collector current. That same action is occurring in Q1; however, this time a PNP transistor is used. To activate a PNP transistor the base voltage must be negative with respect to the emitter, the collector voltage is negative with respect to the base, and current flows from emitter to collector. If point A of *Figure 5-23* is at 0 volts, then Q1 conducts current and the siren sounds. Conversely, if point A is at +12 volts, the base-emitter voltage is zero, Q1 does not conduct and the siren is off.

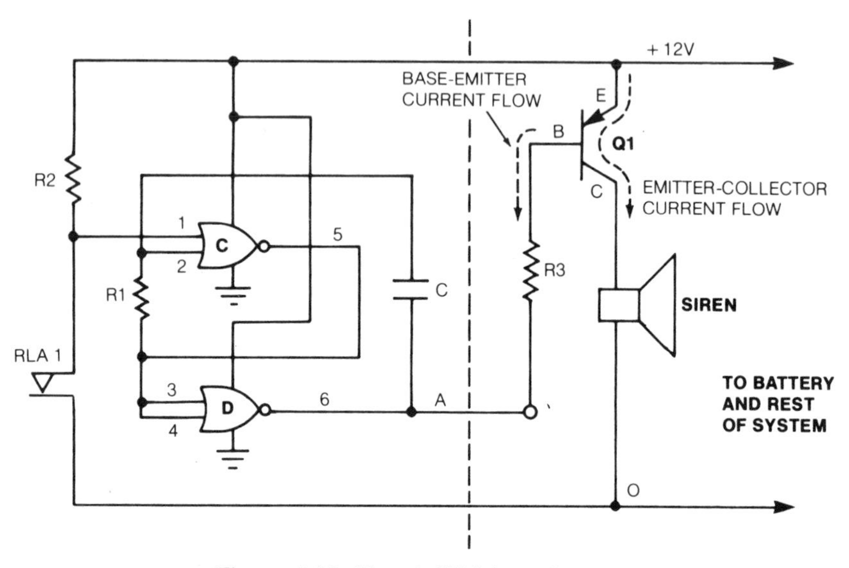

Figure 5-23. *Circuit Which Activates A Siren*

To cause the siren to sound, the voltage at point A is changed rapidly from 0 to 12 volts at the required frequency (800Hz to 1200Hz). *Figure 5-24* shows how this voltage looks when it is plotted against time. This waveform is called a square wave because X and Y are approximately equal. During time X when the level of point A is low (0 volts) the siren is on; when point A is high (+12 volts) during time Y the siren is off. Let's examine how this waveform is generated by starting with *Figure 5-25*.

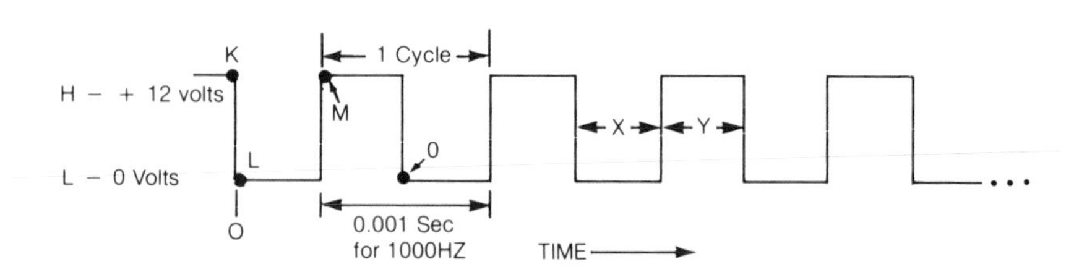

Figure 5-24. *Voltage At Point A Driving Siren*

In *Figure 5-25*, when RLA1 is open and the alarm is not sounding, NOR gate C input 1 is high (+12 volts). This means the output (5) will be low regardless of the condition of input 2. When output 5 is low, inputs 3 and 4 of NOR gate D are also low and its output (6) is high. Therefore, point A is high (+12 volts). A current flows for a short time through the path shown in *Figure 5-25* because the capacitor C charges to the +12 volt level at point 6. This charging current flow causes input 2 to become high while it flows, but as soon as the capacitor charges, the current stops and input 2 returns to the low condition of output 5. The fluctuation of the signal at input 2 has no effect on the circuit operation at this time because input 1 is held at the high level since RLA1 is open. This is at point K on the wave form of *Figure 5-24*.

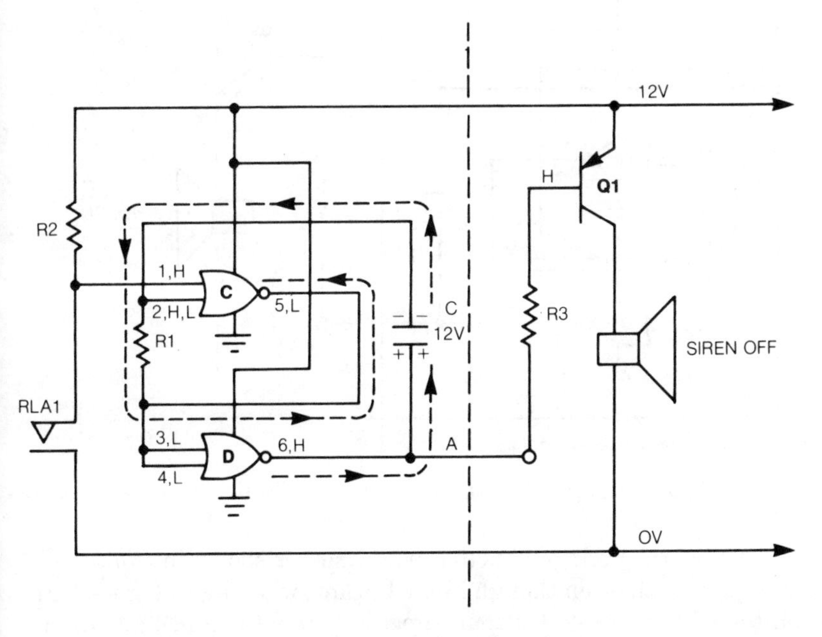

Figure 5-25. *Initial Condition Of Point A At High–Siren Off*

When an alarm condition exists (point L of *Figure 5-24)*, RLA1 closes and pulls the level of input 1 low (0 volts) as shown in *Figure 5-26*. Both inputs 1 and 2 are low so output 5 goes high. This causes inputs 3 and 4 to go high, so output 6 goes low. When output 6 goes low, point A is low and it turns on Q1 and the siren. At the instant that RLA1 contacts close, the capacitor is charged to 12 volts with the polarity shown in *Figure 5-25*. Since output 6 goes low to 0 volts, input 2 will go to −12 volts. The capacitor C then discharges through the path shown and charges to the polarity shown in *Figure 5-26* until input 2 gets to a high level.

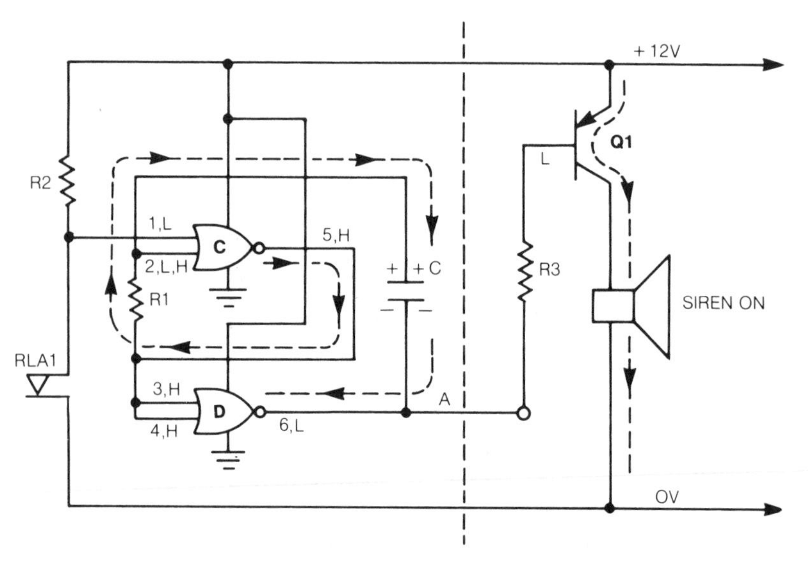

Figure 5-26. *Point A At Low–Alarm Activated Siren On*

Now the cycle begins to repeat itself as shown in *Figure 5-27* and *Figure 5-24*. Even though input 1 is low, when input 2 goes high (point M of *Figure 5-24)* output 5 goes low. Inputs 3 and 4 go low and output 6 goes high, point A is high and Q1 and the siren are off. Now that output 6 is high the capacitor begins to charge up again as it did in *Figure 5-25*. When the capacitor is charged and the current stops, input 2 again becomes low and another transition in the level at point A occurs (point O of *Figure 5-24)*. Due to the waveform at point A, the siren experiences a step function of current and voltage the same as *Figure 5-24*. The values of R1 and C determine the frequency of the signal. The step function of voltage to the siren creates a loud constant noise when connected to the siren. Other circuits of this type can be used to generate "warble" tones – a fluctuating sound. Using various values of R1 and C for the circuits to drive the siren can create different-sounding alarms for different types of emergencies.

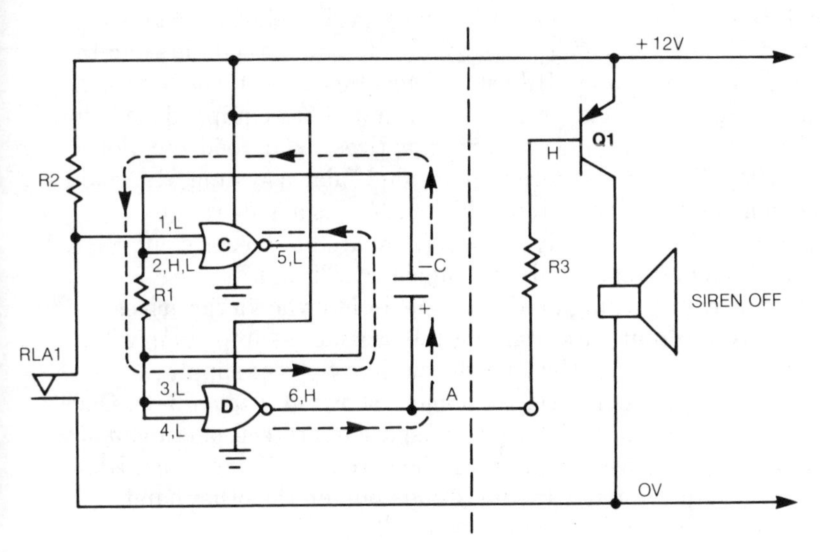

Figure 5-27. Point A When Input 2 Goes High

USING A COMPUTER

System requirements, especially for business and industrial detection systems, may be quite extensive calling for multiple control boxes and very complicated systems. Instead of adding multiple control boxes a computer can be used. This gives an electronic control center which can monitor and control a sequence of changing lights, delayed alarms, sirens with different signals for different situations, remote dialers sending several different messages and a number of both normally-open and normally-closed loops. A significant advantage of using a computer system is that it can handle the security system and still leave much of the computer system free to handle many other tasks such as household finances, regulating the home temperature, keeping track of the need for periodic maintenance, education and entertainment. Let's look at such a complex system installed in a house.

A COMPLEX CASE

Picture the following situation. Dave Sloan and his family have just moved to a midwest countryside. The location is quite isolated and that has increased the need for security, but in addition, it also has brought up the problem of self-sufficiency. To obtain self-sufficiency, at least temporarily, Dave has installed a backup gasoline-powered generator sufficient to power the lights and the basement sump pump. He's stored several cords of wood to use as a backup for home heating in the winter and has equipped his home with a tractor snow plow, extra snow tires, extra food, etc. For security, Dave needs more than a simple alarm system. He would like a home security system which performs a wide range of functions not necessarily all directly related to security but related as well to emergency and self-sufficiency. Things like – automatically turning on the outside lights when a car drives through the front gate, starting the basement sump pump when water collects, and activating the auxillary power supply.

A floor plan of Dave's home is shown in *Figure 5-28*. Dave's home has more than 3500 sq. ft. of space to protect, plus about 450 square feet of glass surrounding the porch and living room. His security requirements are not unique but, on the other hand, demand more than a simple alarm system.

Design specifications

Here are some design specifications that Dave has written down for the system:

1. Certain parts of the system are to be energized at all times. These are:
    a. The freezer alarm – freezer failure is to trigger a special "beeper" type of alarm.
    b. All fire sensors (smoke and heat) – the fire alarm is to trigger a loud siren alarm and then, one minute later, start a remote dialer which summons the nearest fire department (10 miles away).
    c. The circuitry necessary to start and stop the auxiliary generator and the sump pump.
    d. A seismic detector in the driveway – the detector circuit is to turn on outside flood lights. It has a photocell which keeps the circuit turned off during the day.
    e. The time of each of these occurrances is to be stored for future reference.

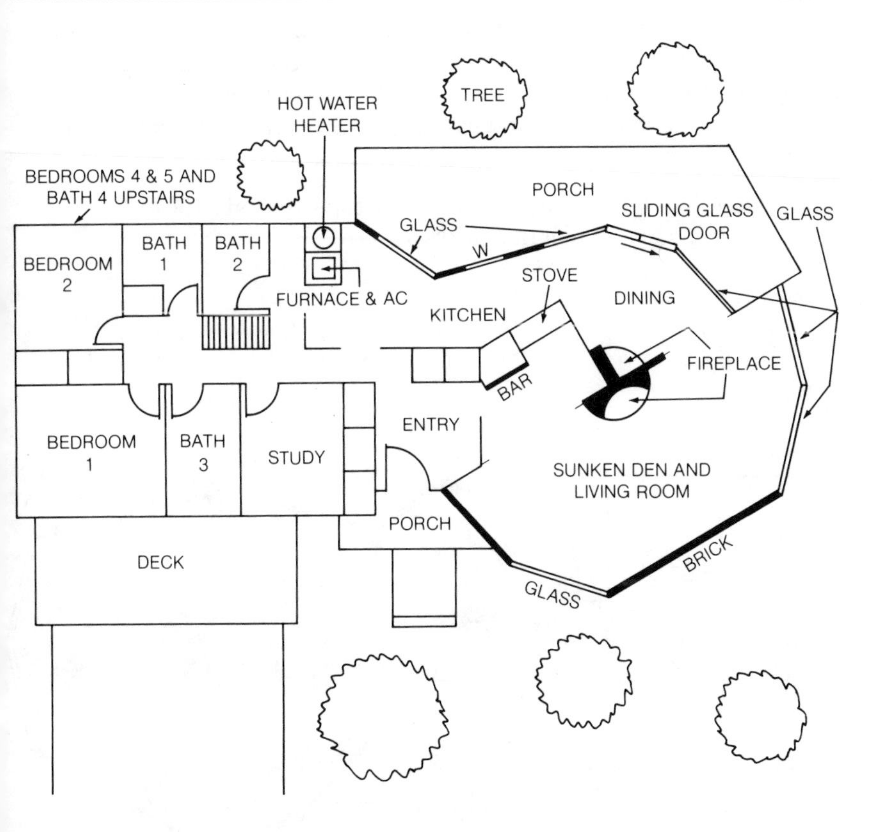

Figure 5-28. *Dave's Home*

2. The outside windows are to be protected by infrared and magnetic switch perimeter loops which may be energized separately. A violation triggers a buzzer and after one minute (if appropriate action is not taken) sounds a loud siren alarm and calls the police via a remote dialer.
3. The inside of the home is to be protected by both microwave and ultrasonic space detectors which work in conjunction with the perimeter system when activated.
4. When both the internal space protection and the perimeter loops are activated, a programmed sequence which simulates an occupied home is to begin. The lights, radio, and a tape player (which plays normal home sounds) are to produce a convincing sequence of events.

5. When a violation triggers a local alarm and initiates the remote dialer sequence it also switches the tape recorder to another channel to monitor the sounds at selected locations in the home.
6. The system should provide exit and entry time delays and panic button inputs when the security system is activated.

The detection devices Dave plans to install to satisfy these specifications are shown in *Figure 5-29*.

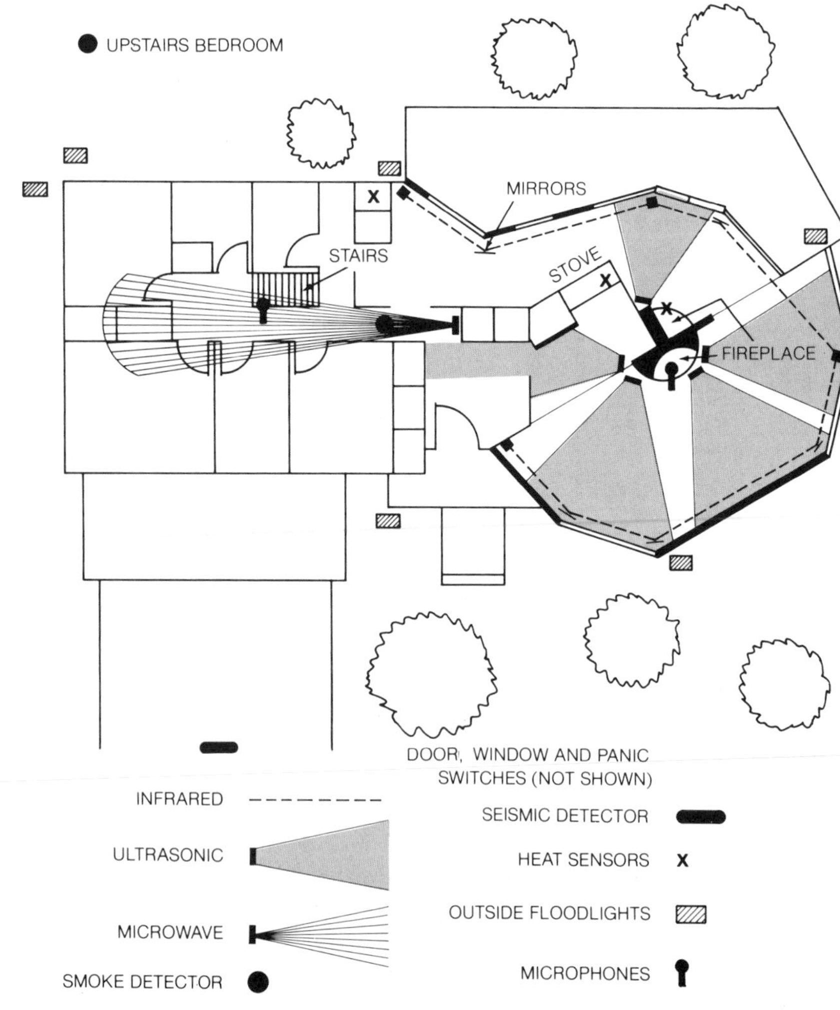

Figure 5-29. *Location Of Detectors*

The Basic System

The heart of the detection system is a home computer. It will silently scan all of the input conditions for any irregularities, any *detection* of the need for action. When one occurs, a response action is stored in its program and it *communicates* and provides outputs that *act*. The alarm is sounded, the lights are turned on, the police are alerted, the fire department is notified, the pump is started, all in response to a particular input signal that is activated.

The necessary sequence of time delays are provided by the computer when the security system is activated to accomplish the exit time delay, the entry time delay, any special turn-on sequences required due to overlapping systems and limiting the time the alarm is sounded for each violation. Certain protective operating time limits may also need to be provided for the tape recorders after they are initiated.

Needless to say the circuitry required to accomplish this task is considerable. Interface circuits that couple the detection devices to the computer are required. A wide variety of these are available but a number of them need to be designed for the particular application. Interface circuits that couple the computer outputs to useable action also are required. A number of these are available, others must be designed especially for the application.

Only a small portion of time is required for the computer to monitor and act as the detection system. The remaining time can be devoted to useful tasks that already have been mentioned.

A block diagram of the system is shown in *Figure 5-30*. Dave plans to do a lot of the interface circuit design himself. He will also write the computer program so this will save much of the normal system expense if he contracted to have a turn-key job done by an outside firm. Even so, the system expense will amount to several thousand dollars. Obviously, he is getting more for his money than just a security system. He is getting a system that he can use extensively as a tool for keeping records, help with his investments, manage his home heating, keep his appointment schedules, remember the family commitments, keep a birthday record, figure recipes, play games, help the family with basic skills in mathematics, language and science and many other applications that will be apparent after the system is completely installed.

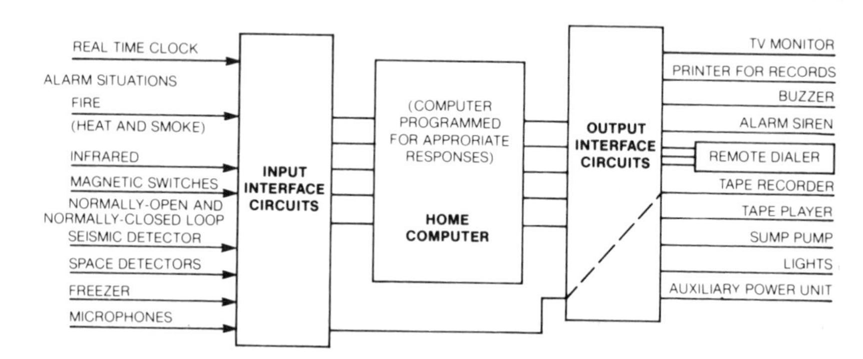

Figure 5-30. *Block Diagram Of System*

One doesn't know how extensively such systems will be used or if our need for electronic detection systems will continue to expand. What is assured is that electronic technology can be applied to aid in keeping us secure. We hope we have provided the understanding of how this is being done.

Quiz for Chapter 5

1. The control box of a modern alarm system often:
   - **a.** Uses both normally-open and normally-closed loops.
   - **b.** Provides entry and exit delays.
   - **c.** Latches so the alarm sounds even after the violation has ceased.
   - **d.** Allows the use of lightweight switches.
   - **e.** All of the above.
2. The relay:
   - **a.** Often uses a small current to switch larger currents.
   - **b.** Is based on the principle of an electromagnet.
   - **c.** Is often replaced by solid-state components.
   - **d.** a and b.
   - **e.** a, b and c above.
3. The time delay circuit discussed in this chapter charges a:
   - **a.** Sweep circuit.
   - **b.** Transistor.
   - **c.** Resistor.
   - **d.** Capacitor to turn on a transistor.
4. The solid-state device used to change alternating current to direct current is:
   - **a.** A transistor.
   - **b.** A diode.
   - **c.** A NAND gate.
   - **d.** A capacitor.
5. The output of a 2-Input NOR gate is high if:
   - **a.** Both inputs are high.
   - **b.** Only one input is high.
   - **c.** Both inputs are low.
   - **d.** None of the above.

6. As an alarm grows in complexity, the control box is often replaced by:
   a. A computer.
   b. Several control boxes.
   c. Several remote dialers.
   d. A series of resistors.
7. Many times the most difficult part of using a computer in an alarm circuit is:
   a. Writing the program.
   b. Interfacing the computer to the detect and act functions of the system.
   c. Locating the computer.
   d. Connecting the wires.
8. Relays are often replaced by solid-state devices which:
   a. Are more reliable.
   b. Require less power.
   c. Are faster.
   d. Occupy less space.
   e. All of the above.
9. Delay circuits are not used in a control box to:
   a. Provide an exit delay.
   b. Provide an entrance delay.
   c. Latch the alarm on.
   d. Turn the alarm off after 5 minutes.
10. A ________ must be used with diodes to convert 110 Vac to 12 Vdc.
   a. Coil
   b. Transistor.
   c. Transformer.
   d. Switch.

Quiz Answers

Chapter 1	Chapter 2	Chapter 3	Chapter 4	Chapter 5
1. f	1. c	1. c	1. b	1. e
2. e	2. e	2. a	2. c	2. e
3. f	3. c	3. d	3. b	3. d
4. c	4. c	4. d	4. c	4. b
5. b	5. b	5. d	5. c	5. c
6. e	6. c	6. e	6. d	6. a
7. e	7. b	7. c	7. a	7. b
8. a	8. a	8. d	8. b	8. e
9. d	9. e	9. c	9. a	9. c
10. a	10. b	10. b	10. d	10. c
11. c		11. d		
12. b		12. d		
13. c		13. c		
14. b		14. b		
15. c		15. c		
16. d		16. a		
17. a		17. c		
18. a		18. d		
		19. a		
		20. c		

Index